1/14/13
$13.95

Praise for
The Dolphin in the Mirror

"Instead of searching the galaxy for intelligent life, we have it right here at home in the large-brained cetaceans. One can wish for no better guide to these fascinating animals than Diana Reiss, who has probed their minds all her life, seeing glimpses of remarkable self-awareness. Here she offers a lively account of her career, and hopes for the successful coexistence of dolphins with a humanity that unfortunately is not always humane."

— **Frans de Waal, author of** *The Age of Empathy*

"A wonderful and passionate tour through the career of a scientist who has made groundbreaking discoveries about the minds of dolphins and helped lead the effort to stop the horrific slaughter of dolphins in 'The Cove' in Japan. A must-read for everyone interested in these amazing intelligent animals."

— **Richard C. Connor, author of** *The Lives of Whales and Dolphins: From the American Museum of Natural History*

"A marvel — Diana Reiss takes a scientific and wholly empathetic look into the dolphins' eyes, and finds them looking right back. As you read this book you'll find yourself wondering: Why aren't we all in the water watching dolphins with Reiss?"

— **Alexandra Horowitz, author of** *Inside of a Dog*

"What an extraordinary story! Diana Reiss has discovered that dolphins have self-awareness, and she has gone from that discovery to help lead the fight to save them. The work she describes here with clarity and passion has changed her life. In the same way, *The Dolphin in the Mirror* changes our sense of what it means to be alive. A new planet swims into our ken."

— **Jonathan Weiner, Pulitzer Prize–winning author of** ***The Beak of the Finch***

"Reiss has managed no small feat — synthesizing personal experience, descriptive material, and scientific fact to provide insights about dolphin behavior for the general public and scientific community alike. Throughout the book, by explaining her 'enchantment by dolphins,' Reiss likewise enchants her audience, ending with an extraordinarily powerful plea for dolphin conservation. No one reading this book could possibly remain untouched by the beauty and intelligence of these powerful mammals of the sea."

— **Irene Pepperberg, author of *Alex & Me***

"Reiss movingly conveys her deepening relationship with the dolphins, and she documents how, through each step of the process, and with each new generation, there is a tremendous emotional pull built upon the establishment of communication and empathy between our different species . . . Engrossing scientific memoir."

— ***Kirkus Reviews***

The Dolphin in the Mirror

PISCE SVPER CVRVO VECTVS CANTABAT ARION

Arion, the seventh century B.C.E. poet, is rescued from the sea by a dolphin
in this illustration by Albrecht Dürer, ca. 1514.

The DOLPHIN
in the MIRROR

EXPLORING DOLPHIN MINDS
AND SAVING DOLPHIN LIVES

DIANA REISS

MARINER BOOKS
HOUGHTON MIFFLIN HARCOURT
BOSTON   NEW YORK

First Mariner Books edition 2012

Copyright © 2011 by Diana Reiss

For information about permission to reproduce selections from this book, write to Permissions, Houghton Mifflin Harcourt Publishing Company, 215 Park Avenue South, New York, New York 10003.

www.hmhbooks.com

Library of Congress Cataloging-in-Publication Data
Reiss, Diana.
  The dolphin in the mirror: exploring dolphin minds
      and saving dolphin lives / Diana Reiss.
      p. cm.
  Includes bibliographical references and index.
  ISBN 978-0-547-44572-4   ISBN 978-0-547-84461-9 (pbk.)
  1. Dolphins — Psychology. 2. Dolphins — Conservation. I. Title.
  QL737.C432R457 2011
  599.53'15 — dc23
  2011016064

Book design by Melissa Lotfy

Printed in the United States of America
DOC 10 9 8 7 6 5 4 3 2 1

For the dolphins

To my husband, Stuart, &
my daughter, Morgan

List of Video Illustrations

The subjects below, from indicated chapters, can be viewed via streaming video at www.hmhbooks.com/dolphinmirror. Specific links to each one can be found in footnotes in the appropriate location of each chapter.

1. Dolphins using keyboards (chapter 3)
2. Dolphins using a learned whistle to represent an object (chapter 3)
3. Dolphins blowing bubble rings and playing with them (chapter 4)
4. Dolphins watching themselves in a mirror (chapter 5)
5. Dolphin spinning and watching herself (chapter 6)
6. Observing and recording wild dolphins in Bimini (chapter 7)

Contents

The Dolphin in the Mirror

Prologue

SAVING HUMPHREY

IN OCTOBER 1985, millions of people the world over followed the plight of Humphrey the humpback whale, a lost, stray, forty-ton leviathan who accidentally wandered into San Francisco Bay and swam far inland. Humpbacks were migrating south along the Pacific Coast, from Alaska to the warmer waters of Baja, Mexico, Hawaii, and beyond, but Humphrey was in danger of beaching and never making it back to the open ocean. At first, few paid attention. But as the days went by and Humphrey remained trapped, the headlines began to appear.

One chilly afternoon, I was sitting on the edge of the dolphin pool at my research facility at Marine World Africa U.S.A. in Valejo, California, feeding two young bottlenose dolphins, Pan and Delphi, when my assistant got a call. The director of the California Marine Mammal Center (CMMC), the regional marine mammal rescue center, explained to my research assistant that it was urgent that she reach me. My assistant took over the feeding of the dolphins, and with my wet hands covered in fish scales I answered the phone. Peigin Barrett, the center director and a dear friend, was speaking quickly about the forty-five-foot-long humpback whale that had swum under the Golden Gate Bridge nearly two weeks before.

Humpback whales are best known for their hauntingly beautiful songs that can travel great distances in the seas. Although the purpose of the songs remains unclear, researchers believe they have something to do with mating behavior, male-male competition, and perhaps social contact and individual identification. Imagine a population of whales spread out over hundreds of miles of ocean, their identity and relative location broadcast through song; effectively, they form an acoustic network. Humphrey had probably become separated from other humpback whales traveling south, and I wanted to help save him.

I was a science adviser for the Marine Mammal Center. I also helped rescue marine mammals. Injured and stranded dolphins and small whales were brought to our facilities, and my research assistants and I worked with a veterinarian, trainers, and other volunteers in efforts to save them. Now we faced a new challenge: an on-site rescue. Whales had been observed in San Francisco Bay waters before, but they generally made brief, albeit well-publicized, tours and then exited uneventfully. Humphrey had turned unexpectedly and wandered inland, swimming through a series of connected bays and waterways, each one smaller than the last, until he was eighty miles from the open ocean! When Peigin called me, Humphrey was swimming back and forth in the Sacramento River and into ominously small, fingerlike sloughs near the small sleepy town of Rio Vista.

⟡

The previous week, a rescue attempt using underwater whale calls had failed. Some of my colleagues, local marine mammal scientists, had conducted a playback experiment; that is, they'd played recordings of the calls of killer whales, a natural predator of humpback whales, hypothesizing that upon hearing such sounds, Humphrey would quickly depart. But it was no surprise when this approach failed. Previous playback attempts over the years using predator calls had failed to deter dolphins and whales from dangerous areas laced with fishing nets. These animals are pretty smart; apparently, they check out their environment, realize there is no true threat, and ignore the acoustic "scarecrows."

By now, Humphrey had been in both brackish and fresh river water for a week and a half, with little or nothing to eat. The water changed the appearance of his skin. Buoyancy is quite different in fresh water than in salt water, and Humphrey had been forced to expend more energy with less food consumption. The clock was ticking. We had to get him back out to sea.

A military helicopter picked up Peigin and me at San Francisco International airport at five that evening and took us to the Operation Humphrey headquarters, a makeshift control center at a U.S. Coast Guard station near Rio Vista.

We landed in the darkness on the bank of the Sacramento River, and Peigin and I were immediately ushered into the bright fluorescent lights of Operation Humphrey headquarters. A meeting room there was already filled with federal staff from the National Marine Fisheries Service (NMFS), the National Oceanic and Atmospheric Administration (NOAA), and the U.S. Coast Guard, as well as CMMC staff and some local officials and townspeople.

A rather stiff-necked NMFS agent whom I will call Dave took charge at the front of the room and began the meeting. He reviewed the past week and a half and Humphrey's travels farther and farther from salt water and food. But Dave stunned us when he expressed his overarching concern: If the whale died in the Sacramento River, his rotting carcass could present a health issue. Saving the whale was, it seemed, a secondary issue.

Dave then brought forth and uncovered what looked like a medieval torture device: a barbed round object on a stick. It was a radio tag that he wanted to use to track Humphrey's location. Radio tracking was an excellent idea, but unfortunately the only tag available had to be attached to the whale by embedding the barbs into its blubber and muscle. The CMMC veterinarians and our rescue staff strongly opposed this idea. The whale was already compromised and stressed, and the barbs would only add to his problems. Dave dropped the idea — at least for the time being.

By the end of the meeting we'd arrived at a plan. The next day, with a flotilla of Coast Guard boats, a few riverboats used in the Vietnam War, and a myriad of small private boats owned and manned by local residents of Rio Vista, we would try to find the whale and form a boat barrier to herd Humphrey back to sea.

We arrived at the dock the next morning and Peigin and I were assigned to the lead boat, the *Bootlegger*, used by some of the CMMC staff. It was a small fishing boat owned and operated by a local fisherman, Captain Jack Finneran, who'd kindly donated his time and vessel to help in the rescue. On the boat with us was another researcher who worked with the CMMC, Debbie Glockner-Ferrari, and her husband, Mark, a wildlife photographer. Debbie had been studying humpbacks in Hawaii and could determine the sex of these enormous animals while swimming with them. We set off upriver in search of Humphrey. En route I used a hydrophone (an underwater microphone) to obtain some recordings of normal noise levels in the river. As we moved northward under the Rio Vista Bridge, I noticed that the noise level was much greater in the waters on the north side of the bridge than on the south side. This finding would play an important role later in the rescue, though I had no inkling of it at the time. Then the boat's radio crackled: Humphrey had been spotted in a small slough near Sacramento. We raced off in the direction of the whale.

I was absolutely stunned to see this huge whale in such a small body of water, flanked on both sides by grassy fields with grazing cows.

Humphrey was an amazingly large yet graceful whale, a lost alien in this bizarre landscape. I could barely see him below the water line until he raised his blowhole out of the water for an explosive breath. We observed him slowly moving through the sloughs; to our surprise and continual frustration, Humphrey demonstrated an uncanny ability to disappear into these very small bodies of water. We tracked him by following his "footprints," smooth, round circles on the water's surface created by his tail movements. Yet at frequent intervals, the footprints would suddenly cease. It was weird; for hours, even aerial surveys couldn't spot him. Our small boats seemed ineffective at guiding him in any direction, no matter how coordinated we tried to be.

At midday I called my colleague Dr. Kenneth Norris; considered by many to be the father of modern marine mammal research, he was the scientist who discovered echolocation in dolphins. A professor at the University of California at Santa Cruz, Ken was not too far away. He joined us for our next meeting at the Operation Humphrey headquarters. Ken urged us to employ a method called *oikomi,* in which a flotilla of small boats is positioned in an arc behind the whale, and then a person on each boat bangs with a hammer on a metal pipe that's partially submerged in the water. This creates a cacophony of syncopated sounds that the whale avoids. The sonic wall moves toward the whale, and the whale is herded forward. Ken provided clear instructions, and we called for small boats, pipes, hammers, and volunteers. Ironically, the *oikomi* technique is used by small groups of fishermen in Japan to herd dolphins to their deaths. For us, it was essential in saving one whale.

◦≪◎≫◦

Soon the dock in the little town of Rio Vista was brimming with local townspeople, CMMC volunteers, and government officials, all of them holding hammers and pipes provided by a local construction company. Local boat owners and fishermen generously volunteered their boats and skills, so we had our flotilla. Ken joined us on the water that day and directed us on how to stay in formation and hammer on our pipes.

We found Humphrey circling slowly even farther north than before. He had passed under a very small overpass, named the Liberty Bridge, made for single vehicles and pedestrians. We stealthily moved north of Humphrey and carefully formed a tight arc behind him. We put our pipes in the water and began to hammer. The sound was like loud underwater wind chimes, a chaotic clamoring that at times created a syncopated rhythm of its own. The small arc of boats moved up behind the whale, and we herded him southward, closer and closer to the Liberty Bridge. The technique worked well, although Humphrey occasionally managed to turn around, slip through a "hole" in our sonic net, and briefly head north again.

As we drew close to the bridge, the whale slowed down. He abruptly stopped within six feet of the bridge's wooden pilings. The pilings were about two feet in diameter and were spaced twelve to fifteen feet apart. Would Humphrey pass through them? He wasn't budging. We moved the *Bootlegger* into a lead position, ahead of the other boats, banged our pipes, and practically rode up onto the whale's tail in an effort to urge him under the bridge. He could easily have brought his enormous eighteen-foot-wide tail down on us hard if he'd wanted to. Humphrey didn't, but he held his ground. He rolled onto his side, raised his huge, fifteen-foot-long pectoral fin, and repeatedly slapped it on the water surface. The Latin name for the humpback whale is *Megaptera novaeangliae; megaptera* translates to "giant-winged." Humpbacks have the longest pectoral fins of all cetaceans. They often lift their fins and slap them on the water surface. The specific purpose of this signal is unknown, but we understood Humphrey that day: he had no intention of moving under the bridge. I watched him slap his fins in obvious agitation and protest and wondered, *What is spooking him?*

We decided that the Coast Guard and NMFS would continue to monitor his movements while the rescue team met to figure out the next steps. As we stood on the riverbank and discussed the situation, I looked back at Humphrey. He was still swimming in the vicinity of the bridge. It was no surprise to me that the whale refused to move through the wooden pilings and under the bridge: marine mammals generally don't like to pass through narrow openings. I had seen this with the dolphins at my lab. We had to acclimate them slowly before they would move through gates or from one pool to another. Man-made passages are unnatural to dolphins and whales. They live in an unobstructed sea.

Yet days before, Humphrey had swum through the pilings heading north. His refusal to do so now couldn't have been due to a lower water level, because we had purposely waited for high tide that day before attempting to herd him through.

I tried to imagine the situation from the whale's point of view. Suddenly, I had a flash of intuition. To this day, I cannot explain it. I just suddenly knew that there was debris — perhaps some old rebar left over from when the bridge was constructed — reaching up like twisted metal fingers from the river bottom. What if the whale had injured himself during his previous passage and didn't wish to repeat the experience? I don't know why I thought this and I know it sounds far-fetched, but as I stood on the riverbank looking at this poor lost whale, I was convinced.

Oftentimes, working with an individual animal, one gains an intuition about the species' general behavior. As suggested by the well-known ethologist Irenäus Eibl-Eibesfeldt, in human and animal interactions, subtle information can be conveyed and interpreted by both sides because "familiarity breeds interpretation." My familiarity with the behavior of dolphins specifically, and whales in general, may have led me to my intuitive glimpse. In any event, it seemed worth exploring. Peigin's eyes lit up at the idea, and she pulled me toward Dave.

With great trepidation we presented the hypothesis and suggested that we check out the river bottom under the bridge with a ship's sonar. Dave said the idea was ridiculous and immediately rejected it. But luckily, at that point California state senator John Garamendi, a tall, elegant, and handsome figure, joined us in the discussion to see if he could help in any way. The senator listened thoughtfully to the idea and agreed that it was worth investigating. A very displeased Dave just shook his head and walked away. The results of a sonar scan proved my hunch correct: old rebar was indeed sticking up from the bottom of the slough under the bridge. That night, a construction crew dredged and removed it.

At eight o'clock the next morning, the small flotilla reassembled and arced the boats to guide Humphrey. We waited for high tide and then tried once again to get the whale to pass under the bridge. This time, Peigin and I were observing the action from the bridge, and I tried to record any vocalizations the whale produced. I didn't want Humphrey to see us on the bridge — it might spook him — so we stayed on its extreme left side, lying on our stomachs on the cool asphalt road. I had my hydrophone dangling below me in the water, and I monitored my recording equipment for sound levels and listened through my headset for vocal signals from Humphrey. Before the din of the *oikomi* banging commenced, I heard a few plaintive-sounding calls from Humphrey. I had no idea what the low-frequency, resonant *hurumph*s meant.

Then I observed Humphrey moving his head from left to right and back again in a scanning motion. I heard what sounded like individual clicks. This was extremely interesting to me because it suggested that humpback whales might use echolocation — biological sonar — to orient themselves, navigate, and detect objects in their environment. (At the time, there had been only one report, by my colleague Hal Whitehead, that suggested the possible use of echolocation by a humpback whale; that case had involved a whale that was trapped in ice.) A subsequent analysis of the clicks at Ken Norris's lab could not confirm that all of the clicks were produced by Humphrey; some of them might have been made by the boats' changing gears. However, some researchers have since suggested that certain whales use low-frequency, repeated sounds as a rudimentary form of echo-ranging. The question still lingers.

It was high tide, and the *oikomi* band began. At first, Humphrey didn't budge, but then he slowly edged forward and stuck his head between the pilings. The boats slowly moved forward behind him. Humphrey proceeded halfway through the pilings and then just stopped. He began to rock his torso from left to right. It appeared that he was stuck, his gigantic pectoral fins wedged close to his sides between the vertical bars. Suddenly, with my stomach pressed to the roadway, I had a sickening feeling. The road below me shook from side to side as Humphrey tried to free himself from the pilings that bound him. Peigin shot me a look as we both imagined the entire bridge collapsing. But we stayed and watched, terrified for Humphrey and for ourselves. Miraculously, with one more shake, Humphrey wiggled free and was through. Humphrey exhaled an explosive blow of rainbowed misty air and then quickly inhaled. I mirrored his behavior in reverse, inhaling deeply and then quickly releasing an explosive breath in relief.

At the Operation Humphrey meeting later that night, all hell broke loose. Dave was clearly agitated. He opened the meeting with an accusing look in my direction. He asked who was responsible for getting someone in the government to turn off the low-frequency coastal monitoring system — a system used to detect enemy vessels — on the theory that the sounds were attracting Humphrey! At the first meeting at Operation Humphrey headquarters, I had heard some rumblings from local residents who believed that Humphrey might have somehow been attracted to or influenced by the coastal acoustic monitoring system. I hadn't taken this concern seriously and therefore was quite shocked to hear that the system had indeed been turned off for a brief interval that morning while we were trying to get Humphrey under the bridge. Apparently, a rumor was circulating that a few members of the rescue team had somehow convinced the powers that be to turn it off. I was surprised to hear that the system had been turned off and also a bit angry; had we known about it, we might have been able to monitor the whale's behavior more closely. But frankly, I never understood why anyone would think that sounds on the coast would affect the whale's behavior inland. I didn't even know where these sounds were being broadcast from. I made it very clear to Dave that I had no involvement whatsoever. Ironically, this event foreshadowed the current concern that midrange sonar may be damaging to marine mammals. In fact, it is quite possible that the navy's sonar monitoring system harms whales, but at the time, the idea seemed far-fetched.

Back on the water, our sonorous fleet continued to herd Humphrey seaward through a succession of increasingly larger and more formidable bridges and possible barriers. With each bridge we faced new challenges. The next hurdle on our southward route was the Rio Vista Bridge, a much larger—half a mile long—steel expansion bridge that spanned the Sacramento River at the small town of Rio Vista. Before I'd joined the rescue operation, rescuers had tried and failed to move Humphrey back southward under it.

It was late afternoon when we approached the bridge. Two small roads flanked the river, and as we drew closer to the bridge, our arced fleet gripped tightly around Humphrey, I noticed a line of cars and trucks stopped on each side. It looked like people were waiting at the finish line of a great race; children were sitting on their parents' shoulders, and people were standing on the roofs of their vehicles, cheering Humphrey on.

And then it happened. Humphrey stopped within feet of the bridge and refused to move any farther. Ours was still the lead boat, and we gingerly maneuvered the *Bootlegger* and the other small boats around the whale and gently but firmly tried to nudge and encourage him under the bridge, but Humphrey held his ground.

And then Dave took control. He called us on our shipboard radio and told us he was coming aboard. He quickly approached the *Bootlegger* in a small Coast Guard skiff and boarded, carrying a small case. Without any discussion, he opened the case, pulled out a dark roundish object, pulled a pin from it, and hurled it toward

Humphrey. I watched in disbelief as the object flew through the air as if in slow motion. It was a seal bomb, an explosive device that's often used in construction sites to clear the waters of unwanted marine mammals. It hit about ten feet behind the whale, sank, and detonated.

Within seconds, Humphrey began twisting his huge body; he made a sudden turn away from the bridge and swam right past us, going north, then promptly beached himself in two feet of water. So now we had a beached whale sixty miles inland!

Our rescue group from the CMMC couldn't believe what had occurred. Some key members of our team exploded in anger and quit the rescue immediately. Peigin and I were equally astonished and angry, but we felt we could not quit. We had a forty-ton whale stranded on the riverbank, and if we didn't do something fast, the physical forces acting on him would soon result in irreversible physiological damage that could kill him. This is a real danger when large-bodied whales become stranded or beached. They often have to be euthanized if they are out of the water too long.

We needed to help Humphrey survive the few hours remaining until high tide would set him afloat again. We had to find a way to keep Humphrey's entire body wet, or his skin would dry out and become damaged. I called the local fire department and asked if they could get fireboats on the river and keep the whale under a fine spray of water. Miraculously, they arrived in minutes. I watched from the bridge, and what a surreal image it was: a whale on the riverbank, with arcs of water over him, rather than arcs of boats surrounding him, saving his life.

Many of the CMMC staff were with Humphrey on the riverbank, some trying to calm him, others digging away the earth beneath him to try to get him afloat. I walked quickly along the small roadway past all the stopped vehicles and through the crowds to get to Humphrey. I was amazed at his size and presence out of the water. His forty-five-foot-long body dwarfed mine. I pressed my hand gently against his skin. It was warm and soft, like the skin of the dolphins I was so familiar with at my lab. I walked farther along and looked into his eye. I had never been this close to a humpback and certainly had never had the opportunity to look one in the eye. But now, despite our two species' ninety-five million years of divergent evolution, I felt a familiarity I hadn't expected, a pattern that connected me to him. His eye was warm and dark purplish brown, rimmed in white like ours, and he followed my movements as I walked near him. I wanted to find some way to let him know we were trying to save him — if only we had some means of communicating. But all I could do was be there with him.

I tried once again to imagine being this whale, to see the situation from his point of view. The noise levels under the Liberty Bridge were quite low, but the noise under the Rio Vista Bridge was another story. I had examined sonograms — sound pictures — of the noise levels in the waters both north and south of the bridge. It was clear there was a dreadful din just under this bridge, probably created by the traffic passing over it and somehow magnified by the metal bridge itself acting as a resonator and projecting the sound into the water. Much of the sound was low in frequency — right in the sensitive hearing range of humpbacks. It hit me: the noise from the bridge was stopping him.

Perhaps when the whale had swum under the bridge before there was less traffic and thus less noise. But now it was close to 5:00 P.M. and the bridge was packed with heavy two-way traffic. My idea was simple: To move Humphrey forward, we had to remove that wall of sound in front of him. We had to stop the traffic.

I shared my idea with Peigin, who immediately agreed that we should talk to Dave and get him to shut down the bridge for thirty minutes at the height of rush hour. Dave listened politely but then quickly vetoed the idea. It was approaching five o'clock on a weekday. There was no way he would consider creating a massive traffic jam. End of conversation.

Peigin and I were not about to give up that easily. We noticed that Senator Garamendi was standing nearby, so again we pleaded our case to him. We showed the sonograms of the bridge noise and explained how both acoustic and physical objects could be perceived as barriers by whales. He said, "I get it." He'd once been a cattle rancher, and cattle were the same way — they didn't like barriers and didn't like to pass through narrow openings. "Let's try it," he said. "I will close down the bridge."

And from that moment on, it all worked. The senator ordered the bridge closed at high tide, and Humphrey squirmed and pushed himself off the riverbank. Our flotilla surrounded him from behind, banged our pipes, and then watched as Humphrey swam right under the bridge.

∽∞∾

Over the following days we continued to move Humphrey southward into larger bodies of connected waterways and into the wider and deeper expanses of the Sacramento River. Many bays dwarfed our small flotilla. It became all too clear that our arc of boats was too small. We needed more boats and we needed bigger boats.

The government sent us several more military river-patrol boats, thirty-foot-long rigid-sided vessels that we referred to as Vietnam riverboats because they'd been used to patrol rivers in the Vietnam War in the late 1960s. These versatile boats had fiberglass hulls and water-jet drive, enabling them to pivot sharply, reverse direction, come to a complete stop from full speed in just a few boat lengths, and operate in shallow, weed-choked rivers. They had been perfect for Vietnam, and they were now perfect for us.

The Coast Guard sent in a very large Coast Guard vessel that occasionally served as our mobile headquarters during the week-long rescue. Our flotilla had grown into a strange constellation on the water. But even with these larger ships, we were unable to keep

Humphrey from slipping through our lines. Each night we strategically positioned the military boats to block the openings of the many connecting waterways and sloughs that led back north in hopes of not losing ground and keeping Humphrey locked in position until we could commence rescue operations again in the early morning. Yet he had the uncanny ability to just disappear, vanish from sight, leaving no trace of his blows or watery footprints. We needed an even larger boat with sophisticated side-scan sonar to track our disappearing Houdini.

I called my professional colleague and friend Terry Kelly, then head of a division of the U.S. Geological Survey (USGS) in Redwood City, California. Terry had kindly provided assistance when I was building my lab, lending or giving me surplus hydrophones and other equipment that I couldn't have afforded otherwise. Like most of the San Francisco Bay area residents, he had been following the story of Humphrey. Terry knew what was coming when I called. I remember hearing him say to his staff, while he was still on the phone with me, "Okay, you guys, who wants to go help save a whale?" The next day, a massive USGS ship equipped with high-tech side-scan sonar and manned by an enthusiastic crew moved into position. I looked back at our flotilla from the deck of the *Bootlegger* and saw what appeared to be a moving city on the sea.

On the USGS ship was a room jammed with equipment for underwater surveying, including large screens that displayed changing images of the waters and terrain we were moving through. I was usually on the *Bootlegger* during the rescue operation, but occasionally I went aboard the USGS ship and watched on the side-scan sonar screen as a fluorescent green phantom Humphrey moved ahead of us. I wondered if this image was at all similar to what dolphins and other toothed whales perceived with their biological sonar as they swam through turgid and murky waters.

But even this advanced technology failed us when we tried to track Humphrey through the night. As before, every evening he would somehow ditch us, disappear off the sonar screens, and we'd have to wait for the aerial survey of the morning to locate him again. But even with our enlarged flotilla and advanced technological prowess, we could not control Humphrey's movements in these larger waters. We needed another plan. Senator Garamendi arranged a meeting in the state capitol building in Sacramento.

I arrived at the capitol building that morning and sat down at a very large conference table in what I can only describe as a warroom setting. Senator Garamendi was officiating. With us were Peigin Barrett; Laurie Gage, our veterinarian; other staff members from the CMMC; Bernie Krause, a musician and bioacoustician who had assisted in the rescue; Dave and a few NMFS staffers; and people from NOAA. Present via teleconference were several internationally known marine mammal scientists and colleagues. Among them were Dr. Ken Norris; Karen Pryor, a respected dolphin behaviorist now known for her groundbreaking clicker method for training dogs and other animals; and Dr. Louis Herman, a cognitive psychologist at the University of Hawaii and director of the Kewalo Basin Marine Mammal Laboratory.

At first we were all stumped about how best to proceed. Then Lou Herman made the brilliant suggestion of trying to attract Humphrey to the boats by broadcasting, or playing back, sounds of humpback whales. Previous attempts to attract whales using playback had met with little success. However, after a brief discussion of past failures, we realized we had a unique situation with an isolated whale, so we decided to try it. Lou said he would give us his lab's recordings of humpback whales feeding in Alaskan waters. The government would send a special transport plane to Hawaii to get the tapes and then deliver them to us in California the next morning.

Then, before I even had time to think about it, I was put in charge of doing the playback. I had never done a playback experiment; I hadn't even had much experience with humpbacks before Humphrey. I was flying blind, somewhat terrified, but I agreed to do it since I was the only acoustician and behavioral scientist on the team who worked directly with marine mammals. None of my colleagues had any helpful suggestions about how to conduct the playback, so I was pretty much on my own.

⚬◆⚬

The technical arrangements were made easily enough. That wasn't the issue. Acoustics experts at the Naval Postgraduate School in Monterey, California, sent a few of their grad students with a special underwater speaker we would use to broadcast the whale calls from the *Bootlegger*. But I puzzled over what method to use — how exactly to do the playback.

I went to my lab and sat by the dolphin pools to ponder it all. I watched Circe and her calf Delphi and Terry and her calf Pan as they interacted, listening to their vocalizations over my headset. In my years of watching dolphins in my lab, I'd been struck by how quiet dolphins were when they were close together. At times, the pools would be filled with a diversity of whistles, squawks, echolocation clicks, and other sounds that humans have yet to decode, but there were often long stretches of complete silence when the dolphins were interacting or in mother-calf pairs swimming closely together. As I continued to watch the dolphins, I kept thinking about the playback. Wouldn't Humphrey just check out the sounds, recognize their falsity, and leave? Would he ignore or just get bored with the sounds? How could we use the sounds to keep him with us?

And then, the dolphins inspired an idea. I had observed it so often; it was so simple. Dolphins often produce contact calls when they are separated from each other, and then one or the other re-approaches, and they come together. I had seen it used among larger groups of adult dolphins, between Circe and her calf Delphi, and between Terry and her calf Pan day in and day out. That was it! I would use the playback calls like contact calls; I'd broadcast them to Humphrey when he wasn't near us, and stop broadcasting them when he was with us. It was a hunch, and frankly, it was the only idea I had at the time.

At six in the morning on November 3, I arrived at the dock where the NMFS officials and the CMMC staff were assembling for the event. The *Bootlegger* and a few smaller Coast Guard Zodiacs were already there. The students from the Naval Postgraduate School were making the final adjustments to the speaker system they had installed the day before. The other boats were out of sight, waiting upriver. I had requested that the rest of the flotilla be at least five hundred to a thousand yards behind the *Bootlegger* in order to give us and Humphrey a little personal space. Humphrey's location was unknown, despite the aerial searches to find him.

❦

I motored upriver in a Zodiac with my monitoring equipment to test and adjust the broadcast level of the sounds to be used in the playback. After dropping my hydrophone in the water, I radioed the okay back to the *Bootlegger* to broadcast the test sounds. As I heard the haunting whoops and deep *hurumph* signals of the feeding humpbacks, I noticed a whoosh of water and movement. Humphrey had appeared out of nowhere and was making a beeline toward the *Bootlegger*! We raced past him in the Zodiac, got to the *Bootlegger*, and jumped aboard. Operation Playback had begun.

I stood at the stern of the boat, my eyes fixed on Humphrey as we moved slowly away from the dock and into the deeper waters of San Pablo Bay. Humphrey followed us like a lost puppy. When he was with us, he stayed within five to twenty feet of the boat. I watched his subtle movements, reading his body language, looking for any cues that might indicate that he was about to depart. If it seemed like he was staying, I would signal to stop the broadcasting. If Humphrey began to drift away, we would broadcast again, and he would re-approach the boat. I was thrilled: this was actually working! From the upper deck of the *Bootlegger* we recorded how long we played the sounds before the whale approached the boat and how long he stayed with us once we stopped broadcasting. Humphrey's behavior changed noticeably — he went from a whale swimming around aimlessly to one who had a focus. It was as if he had met a long-lost friend.

We moved the whale about forty miles in four hours, and all was going well. Then a brief episode of tension occurred when Dave radioed from one of the boats in the flotilla and asked us to stop and take a short break. His request made no sense at all to any of us on the *Bootlegger;* given our success with the playback and the whale's momentum south, we wanted to keep up the pace. But Dave was in charge, and I was *ordered* to take a break.

The *Bootlegger* slowed down and stopped, and so did Humphrey. It was then that Dave's real intent became clear. Dave approached in a Zodiac holding a stick with the same barbed radio tag we'd seen earlier. He planned to radio-tag Humphrey now that the whale was trapped between Dave's boat and ours. I called out and told Dave to stop or I would leave the *Bootlegger* and not continue the playback.

Suddenly we were in a standoff. I held my ground and stated that in my opinion, the whale saw the boat and heard the sounds we were playing as a positive stimulus — our playback was working. Sneaking up on Humphrey and attempting to radio-tag him with a device that we'd all agreed was not well developed and might only stress him more would jeopardize the operation, the whale, and the people aboard the *Bootlegger*.

It was very tense out there on the water for several minutes, but fortunately the NMFS officials back at Operation Humphrey headquarters radioed us to say that they thought we should not risk losing Humphrey's trust and that we should continue with the original plan.

Finally, Dave agreed to proceed with the playback. He seemed to relax, and he said he wanted to join us. He jumped aboard the

Bootlegger, and we continued to move the boat southward, Humphrey following us closely. Between that day and the next, we traveled sixty miles, heading toward the final gateway to the open ocean, the Golden Gate Bridge.

The majestic Golden Gate Bridge was the site of the most amazing experience in the rescue. We approached in the late afternoon on November 4, just as the fog began rolling in from the sea, blanketing San Francisco. We entered San Francisco Bay, and I looked across it to the pyramid-shaped Transamerica building in the city, its apex sunlit and shining like a welcoming beacon. We were almost home. The *Bootlegger* moved closer to the Golden Gate Bridge, and Humphrey was no longer following us but swimming about twenty to sixty feet off the boat's starboard side. We were not in the lead any longer. Instead, we were accompanying each other. The large flotilla of boats moved in behind us, creating the familiar arc, gently encouraging Humphrey forward. We stopped broadcasting the sounds because Humphrey looked like he was well on his way back home now.

It wasn't until we were close to the Golden Gate Bridge that I looked up and saw that all the traffic had come to a stop. Multitudes of people were standing on the bridge, waving and cheering Humphrey on. The only mechanical sound I heard was the buzzing from swarms of news helicopters overhead.

I asked the boats in the flotilla to stay back and let us go under the bridge alone with Humphrey. Because the waters near the bridge were very choppy, we had to take a few minutes to pull up the underwater speaker. Humphrey stayed beside us, as if waiting.

I looked over at Humphrey next to us as we started to pass under the bridge. The *Bootlegger* moved slowly and passed through, but Humphrey didn't. He was still on the bay side of the bridge — just floating there. So we circled back and positioned the boat next to him. Again we tried, and again we went under the bridge alone. But on the third try, at 4:36 P.M., Humphrey joined us and passed under the bridge as the crowd cheered.

Our flotilla followed us closely, and once they were all on the other side of the bridge, the other boats quickly encircled the *Bootlegger*. From a bird's-eye view, our small boat must have looked like the bull's-eye in the center of a dartboard. Silently, all of us aboard the boats watched Humphrey swim away, farther and farther westward, finally home.

But it wasn't over. He was swimming *in the wrong direction!* He was heading north, taking him ever farther away from his annual winter migration southward. I asked the boats to remain in the circle formation but to put their motors into neutral. We looked out in all directions for any sign of the whale, but he had once again vanished. We waited five minutes — nothing. Ten minutes — nothing. Then, suddenly, Humphrey reappeared alongside the *Bootlegger*! He had somehow passed unseen under all the other boats and found us. With apparent deliberation, he swam slowly toward our boat. We all waited and watched in silence. He stopped, pressed his belly against the side of our boat, looked up at us for several long seconds, and then swam off, southward this time.

It's hard to describe the force of emotion that I felt as I looked down at Humphrey. Aside from when he was beached, I had seen only his back and his blowhole, a huge dark moving mass behind us. But now he had made contact, body contact and eye contact, that evoked a visceral reaction, a knowing, a real connection that linked us. Two very different species, separated by ninety-five million years of evolution, looking at each other in a way that made a connection. As I write these words I reexperience the raw emotion. How can I explain what it is like to be enchanted by a whale?

༺❀༻

Humphrey was spotted the following year in the Farallon Islands with other humpback whales. In 1990, he was back in San Francisco Bay. This time he beached himself on the bank near Candlestick Park and once again had to be escorted back out to sea. It was during this rescue that it was discovered that Humphrey was actually a female! After that, I waited and hoped for sightings, increasingly worried about this whale with whom I had formed such a strong bond. A few years later I received the very sad news that Humphrey had died. Numerous unanswered questions remained after our encounter with this elusive visitor who moved us in so many ways.

∽∾

In the more than two decades that have passed since Humphrey's misadventure, scientists' understanding and appreciation of the character and capabilities of minds other than our own species' has been dramatically transformed. Prior to the 1950s, it was commonly held that all creatures other than humans were mere unthinking automatons, devoid of intentionality and devoid of any spark of self-awareness. Researchers studying animals' ability to communicate with humans were expected to keep the animals at arm's length, literally and figuratively. Developing a relationship with one's subject of study was unacceptable because it was believed that subtle cues might influence both the animal's behavior and the scientist's interpretation of that behavior. The perils of anthropomorphizing were lurking everywhere, it was said, and were to be avoided. There were still strong echoes of this stance in the 1980s.

Yet at the same time, since the 1960s there has been a growing awareness and concern for the plight of whales. After the discovery in the early 1970s of the hauntingly beautiful songs of the humpback whale, our appreciation of them soared and probably contributed to achieving a successful yet all too brief moratorium on whaling by the International Whaling Commission.

People rally to save individuals. They seem to want to help individuals more than they want to help entire populations. This was the case with Humphrey. And yet, whole populations of humpbacks, made up of individual whales just like Humphrey, are still hunted in many parts of the world. The problem is that the idea of a population is abstract, whereas one or two individuals that we can see and even name become real to us. Real individuals can experience pain, fear, and suffering, and we want to help. I have dedicated my career to understanding dolphins — one species of small whale, that is — and to rescuing them. This book summarizes my life's work, along with the research of others and dolphin lore through the ages, in order to make a bold claim: Dolphins are among the smartest creatures on the planet — fully conscious, creative, and highly communicative, with an intelligence rare in nature. And yet, despite this and the fact that many people and entire cultures have loved and revered dolphins for centuries, mankind is slaughtering dolphins at astonishing rates.

We would never slaughter chimps, and there are laws against slaughtering elephants throughout Africa and Asia. Yet their sentient, empathetic cousins in the ocean are subject to mass killings. My hope is that everyone who reads this book will be motivated to support increased protection of dolphins and whales globally.

1

MINDS IN THE WATER

> The hunting of dolphins is immoral, and that man can no more draw nigh to the gods . . . or touch their altars with clean hands, but pollutes those who share the same roof with him, who willingly devises destruction for the dolphins. For equally with human slaughter the gods abhor the deathly doom of the monarchs of the deep.
>
> —OPPIAN, Greek poet, in Halieutica, approx. 200 C.E.

TEN YEARS BEFORE my close encounter with Humphrey, the idea of studying dolphins (much less whales) was absolutely not in my life plan. As a young child, I had felt very connected to animals, had an innate compassion for them, it's true. I had a dog, Rusty, and a younger brother, Bob, and although I loved my brother dearly I now sheepishly have to admit that in some ways I always felt closer to my dog. I really believed I could hear him thinking and that we shared a very direct form of communication that I didn't have with my brother. I can only imagine now that I was very attuned to Rusty's body language, and my childhood fantasies filled in the rest. One of my earliest childhood memories is of a neighbor tidying up her yard and accidentally disturbing a nest of wild rabbits near our house. I felt it was my job to find all the baby rabbits and the mother, and take care of them. I was a born animal rescuer. Injured birds. Injured frogs. Injured animals of any kind. Little Diana of Assisi was always there to make things right.

Later, I toyed with the idea of becoming a vet, as many kids who love animals do. I also had a talent for art, however, and so after high school I attended Moore College of Art in Philadelphia.

Eventually, I pursued a career in theatrical set design and I began an MFA program in theater and communication at Temple University. It was then that I met, worked with, and married Stuart Firestein, the director of an experimental theater company in Philadelphia where I was a set designer for several years.

But even as I was building theaterscapes, I had a strong pull toward science. One day, I had an epiphany under the most bizarre of circumstances. Stuart and I were participating in an actors' workshop in Poland run by the famed Polish counterculture director Jerzy Grotowski; I'd been quite honored to be invited to take part in it, especially given my limited acting experience. I found myself in a darkened warehouse in Wroclaw, Poland, in the company of about two dozen actors, most of whom spoke no English. As part of an experimental exercise, we were making animal sounds in the dim building. I was thinking to myself, *This is really interesting. I can't speak to these people because I don't know their language and they don't know mine, but we are communicating with one another by making animal calls across the darkness.* I cannot explain it now, nor could I at the time, but I experienced a powerful intuition at that instant, as if a voice were saying to me, *This is not right; you have got to get back to science.* When I told Stuart, "I'm out of here, I want to work with animals," he thought I was completely crazy.*

Crazy or not, I applied to the Speech and Communications Department at Temple University and was accepted into a PhD program in bioacoustics, a cross-disciplinary field that combines biology and acoustics. I had to scramble to take some basic science courses before I could embark on a graduate program, which I was fortunate to be able to fashion for myself around the science of analyzing animal calls, language development, symbolic behavior, animal behavior, cognitive psychology, and communication theory.

---

* Some years later, Stuart made the same "crazy" decision, and he is now a professor of neurobiology at Columbia University in New York.

I wanted to be equipped with skills to understand the communication and behavior of other minds, animal minds. But even then, dolphins were not in the picture for me. That would require one of those chance events many of us experience once in a great while, events of no great inherent significance but that have the effect of changing our lives.

It was a rainy Sunday afternoon in the Germantown section of Philadelphia in the late winter of 1976, perfect for settling down for a couple of hours of serious reading of the *New York Times*. I was now a doctoral student in my first year of graduate school. Stuart and I were living in an old pseudo-Tudor apartment building in a lovely wooded neighborhood. Freezing winter rain was running down the mullioned windows, distorting the images of the trees outside so that they looked like part of an impressionist painting. The apartment was sparsely furnished with props from retired stage sets I had designed. I was sitting at my dark mahogany carved desk, a prop from the show *Mark Twain Tonight!* Prominently displayed in the International News section of the *Times* was an article on the killing of whales and dolphins, accompanied by a big photograph. It was as if I were somehow primed for that moment, because I read every word avidly, turned to Stuart, and said, "It's terrible that these animals are being killed off, and we know so little about them." I wrote much the same sentiment in a diary I kept for only the most important moments in my life.

Public awareness of the precarious plight of many species of whale was growing at the time, and the efforts of the International Whaling Commission were much in the news. Roger Payne and Scott McVay had published a landmark paper in *Science* magazine in 1971 reporting that humpback whales sang songs with very complex structures, similar to classical music pieces, with units, phrases, and themes. They released an album of the humpbacks' eerily moving songs to tremendous interest and acclaim. The beauty of their songs touched me, but I was far from the science of it all. Philadelphia had an aquarium called Aquarama, which had several dolphins on display. I had never visited it growing up and had had no desire to go there. I had no interest in watching dolphins jump through hoops.

I had never been a fan of the *Flipper* television series, which had had a tremendous following in the mid- to late 1960s. The Flipper character was a bottlenose dolphin (played by five female dolphins, and the occasional male for a particular trick) who, the story line went, lived with Porter Ricks, a warden in the fictional Coral Key Park and Marine Preserve in the Florida Keys. Week after week, Flipper helped Ricks protect the park and its wildlife and aided in tracking down and capturing various miscreants who were up to no good there. I disliked the program, thought it was stupid. I preferred *Lassie*.

And yet, when I read the *Times* article that rainy Sunday afternoon, my attention was immediately arrested. As I said, it was as if I had been primed for that moment, but in a way that was not at all obvious to me then and still isn't now. I instantly became ravenous for anything and everything that had been written about dolphins. I scoured the scientific literature and found many papers on dolphin brain anatomy and communication, and I consumed John Lilly's popular-press books on his observations and speculations, *The Mind of the Dolphin, Man and Dolphin,* and others.

Lilly was an American neuroscientist, philosopher, inventor, and writer, a man who delighted in being seen as both a pioneering scientist and a maverick. In the 1960s and early 1970s, he was a member of the California counterculture of scientists, mystics, and thinkers, an occasional acquaintance of the likes of Allen Ginsberg and Timothy Leary, and aware of the psychopharmaceuticals that implied. He was a larger-than-life character, and his research on dolphin minds was driven by a desire to understand consciousness, an ambitious quest that has occupied scientists and philosophers for millennia. His unorthodox approach, and his conviction that the dolphin mind was in some ways quasi-human — that we were destined to communicate and understand each other — elevated dolphins to almost mythical status in the eyes of his followers. There were many of them.

∽∞∽

In addition to doing pioneering scientific work, Lilly single-handedly created a new mythology of dolphins that went far beyond science. In 1975 Lilly put together a compilation of his earlier books and papers in a volume entitled *Lilly on Dolphins — Humans of the Sea*. The concept of humans of the sea may seem a stretch and perhaps anthropomorphic in the extreme, but it was certainly not the first time this idea was put forth. The phrase *humans of the sea* had been bestowed on dolphins by the Maori in New Zealand; John Lilly was merely the latest in a long line of dolphin mythologizers.

Humans and dolphins could hardly be more different in our physical forms and in the worlds we inhabit. And as mammals, our two species could hardly be more distant from each other, being separated by a gulf of ninety-five million years of evolutionary time. We humans are bipedal primates equipped with dexterous hands and guided through a terrestrial environment principally by an adequate, though not superior, visual ability. Dolphins have the hydrodynamic form of fish (no arms, no legs), and they navigate through their aquatic world guided by supersensitive sonar. And yet, from the earliest records of civilization, humans have felt a deep affinity with dolphins.

We see a reverence for these monarchs of the deep in origin myths from Australasia to North America, from Europe to South America, and across Asia. In some of these ancient stories, humans are said to have arisen from dolphins, while in others dolphins are the mythical progeny of humans. Indeed, stories of dolphins being transformed into humans, in origin myths and in other circumstances, are a recurrent theme in cetacean mythology. Arguably, no animal plays a greater role in human mythology and lore than dolphins.

Human esteem for dolphins reached its zenith in ancient Greece, where dolphins were viewed as being closer to the gods than any other creature, half divine themselves, and messengers between the human and divine realms. "Diviner than the Dolphin is nothing yet created," wrote Oppian in 200 C.E., "for indeed they were aforetime men and lived in cities along with mortals." Killing a dolphin in these times was therefore a sacrilege against the gods and was punishable by death. (By contrast, slaves could be killed with impunity.) Images of dolphins — on coins, seals, bronze statues, floor and wall mosaics, and vases — were as much a part of the iconography of Greek culture as marble temples and philosophers in white togas. A common image is of a boy, sometimes resembling Apollo, astride a dolphin and playing a lyre, symbolic of bringing wisdom and the arts of civilization from the sea to the land.

Apollo, one of the more powerful deities in the Greek pantheon, is famous for establishing the oracle of Delphi on Mount Parnassus. The story of how this came about has many versions, as is common in Greek (and Roman) mythology. As the sun god, Apollo was also the epitome of music, poetry, beauty, youthfulness, and grace. Because he loved humanity, he decided he would bestow upon the Earth his wisdom and insights, which would be imparted through his oracle at Delphi.

To this end, one evening Apollo made himself visible to a group of Cretan businessmen sailing in the Gulf of Corinth. He assumed the form of a dolphin, leaped high above their ship, landed on its deck, and changed into the form of a golden youth. He announced to the astonished and fearful group, "Behold, I am Apollo Delphinus!" He told them of his grand plan, and soon the ship's sails filled with wind, the rudder set a new course of its own accord, and the ship surged forth with steadfast purpose. It was clear to the men that there was something greater than mere mortals at work. Apollo resumed the form of a dolphin for the rest of the journey and lay regally shimmering on the deck.

Soon the ship arrived at a port on the southwestern spur of Mount Parnassus. The men disembarked, and Apollo, once again a golden youth, led them to the temple of the oracle of Python, where they were met by Pythia, the chief priestess of a sisterhood that had maintained the oracle for many years. Pythia was displeased at the aggressive intrusion, and she challenged Apollo to a duel. Apollo prevailed, but rather than killing her, as was his right, he honored her for her bravery and for the years she and her sisters had tended the temple. He declared that henceforth, Pythia and her sisters would take on a new role: the voices of the new oracle of Delphi. Inscribed on the walls of the Temple of Apollo at sacred Delphi were words of righteousness and wisdom, the most famous of which was *Know Thyself.*

It is thought that the dolphin god arrived around 1000 B.C.E., and his influence, via the voices of the oracle at Delphi, persisted for almost a millennium and a half, the most powerful political and spiritual presence of the time. There is a link between *Delphi* and *dolphin: Delphi* is the Greek word meaning "womb," nurturing source of life; *delphis* (dolphin) loosely translated means "womb fish" — dolphins, unlike other fish, give birth to their young.

There are several versions of the origin of dolphins in Greek mythology, all of which involve Dionysus, the god of wine and wildness; some are said to have occurred when he was a boy; others when he was an adult. Here is one version.

Dionysus, who had great good looks and grand demeanor, disguised himself as an ordinary traveler and hired a ship and crew to take him from the island of Ikaria to the island of Naxos, the largest island of the Cyclades, in the Aegean Sea. The crew, who were pirates of sorts, believed Dionysus to be a prince and so plotted to kidnap him and profit by some means. They sailed the ship past Naxos and on toward Asia. Dionysus realized that he had been tricked, and he used his divine powers: the masts sprouted branches, the men's oars became snakes, and strange flute music sounded throughout the ship. The men realized that their captive was in fact a god with terrible powers, and they flung themselves into the sea. Poseidon, god of the sea, promptly changed them into dolphins and ordered them to be servants of mankind forever and exemplars of virtue and kindness.*

The first to benefit from the newly created dolphins' selflessness was Poseidon himself, even though he was a god and not of mankind. Poseidon was in pursuit of the beautiful sea goddess Amphitrite, but she was being coy and hid from her pursuer in a cavern under the sea. Dolphins discovered the location of the reluctant bride and told Poseidon where she was hiding. He found her and took her for his wife. To show his gratitude, Poseidon conferred upon dolphins the highest of honors: he created the constellation of the dolphin, Delphinus, which can be seen in the northern skies close to the celestial equator.

Although classical Greece saw the height of human reverence for dolphins in origin mythology, the sentiment has a very long history. Paintings and engravings in prehistoric caves in Europe have long intrigued modern scholars, although their exact meaning will, of necessity, remain elusive. Most agree, however, that Paleolithic people were not simply making visual and tactile records of the various animals of the day — horses, bison, bears, mastodons, and so on. Rather, the depictions likely held some symbolic value, perhaps a kind of ritual relating to the hunt or an encapsulation of their cosmology, their origin myths. It is therefore significant that in the Nerja caves of southern Spain, deep in a barely accessible corner, there are images of three dolphins, two males and one female. And engravings of dolphins are said to be in Ice Age caves in the French Pyrenees. Such images aren't common, yet that they exist at all is remarkable. Exactly what they mean, we cannot know. They are

---

* Plutarch, Greek moralist, recognized this selflessness in his treatise *On the Intelligence of Animals*, 66 C.E., saying, "To the dolphin alone, beyond all other, nature has granted what the best philosophers seek: friendship for no advantage."

tantalizing threads of evidence that mankind's close identification with dolphins stretches back ten thousand, twenty thousand years, and possibly more.

But we can know some of what was in the minds of a people in a different part of the world many thousands of years ago, because their stories have been passed on through countless generations. These people are the Australian Aborigines, a highly diverse group living over vast territories and whose history goes back perhaps fifty millennia. Throughout their diversity is one commonality: a reverence for dolphins, for their sacredness, their wisdom, their spiritual guidance. This special connection between humans and dolphins among Australian Aborigines may well be the oldest one of all of human societies. Stories of dolphins are an integral part of the Aborigines' Dreamtime — that is, the time of the creation of the world in Aboriginal mythology.

Here is just one example. The Wanungamulangwa people live on Groote Eylandt, off the north coast of Australia, and their ancestry goes deep into Dreamtime. Their earliest forebears were said to be dolphins, the Indjebena, who lived in the deep waters between the islands of Chasm and Groote. At that time, the Earth was inhabited by spirit beings in the form of animals, birds, and fish. In the stories of Dreamtime, the Indjebena had a carefree and joyful life, with plenty to eat and plenty of time to play.

Dinginjabana, the leader of the dolphins, was swift, bold, and, it has to be said, more than a little arrogant. His wife, Ganadja, by contrast, was timid and kind. Ganadja was friendly with the Yakuna, a type of shellfish that built a strong shell and had a single muscular foot. In what is a long and quite complicated story, Dinginjabana exhorted his fellow male dolphins to sport with the Yakuna, and they tossed them around with disdain and derision, taunting them for having to stay in the coral, unable to move swiftly like the Indjebena. A mistake, as it turned out, because the Yakuna had powerful friends — the tiger sharks, deadly enemy of the Indjebena.

Baringgwa, the leader of the Yakuna, called upon his shark friends for help, and before long every one of the Indjebena had been sliced and mangled in their ferocious jaws. Every one, that is, but Ganadja; she was given refuge by her friends the Yakuna, who shielded her with their hard shells. After many months of loneliness, Ganadja gave birth to a son, whom she named Dinginjabana, after his father. He grew much larger than his forebears and was no longer at the mercy of the tiger sharks. The young Dinginjabana was the first of the tribe of dolphins that thrived around Groote Eylandt and in the world's oceans, the dolphins we see today.

According to the stories of Dreamtime, the souls of Dinginjabana the elder and the rest of the Indjebena became hard and dry, and after many years they were reborn as humans on Groote Eylandt, the first humans in the world. Meanwhile, Ganadja lovingly raised her son but remained lonely and missed her errant husband. One night, under a full moon, Ganadja swam near the shore and saw her husband, who was now a two-legged man. Overcome with excitement and longing, she thrust herself ashore, dragged herself over the sands with her flippers, and rested in front of Dinginjabana. When he recognized his wife, he gave a great shout of joy; Ganadja joined him in voicing elation, and she promptly took on human form. Ganadja and Dinginjabana lived a very long time and produced many children, who populated the island of Groote. They are the only ones who remember that dolphins are the ancestors of the entire human race. However, the dolphins in all the oceans, the offspring and descendants of the great mother Ganadja, have never forgotten that the people of Groote are their two-legged cousins. That is why, they say, dolphins are so eager to approach and play with their human kin, as they did in the days of Dreamtime.[1]

The Maoris, the Aboriginal people of New Zealand and geographical neighbors of the Australian Aborigines, also have a long and sacred relationship with dolphins. To the Maoris, dolphins are a source of spiritual guidance and a font of wisdom in difficult times. Dolphins, in these people's world, are known as humans of the sea.

On the other side of the globe, the Chumash Indians of the south California coast tell a different story of their origins and the origin of dolphins. Hutash, the earth goddess, lived on the island of Limuw (known today as Santa Cruz Island), where she talked to the animals and the trees, which she cherished. But she was lonely and wanted other people to be with, to share with her the beauty of her beloved Limuw. So one day she climbed the highest mountain of Limuw, gathered poppy seeds, and strewed them over the land. The seeds germinated and matured and grew into men and women, young and old. These were the Chumash people, whom Hutash loved as her own. Hutash's husband, the Sky Snake (the Milky Way), gave the Chumash fire, and they thrived and multiplied on their beautiful island.

Before many years had passed, Limuw had become crowded and the Chumash too boisterous for Hutash's liking. She told them that half of their people must leave for the mainland, and that in three days she would construct a bridge for them that would go from the highest mountain on Limuw to the highest mountain across the water. She warned them that when they crossed it, they must not look down. When the third day came, the families that had elected to leave set out across the beautiful arc of colors that Hutash had constructed. Very soon, some of them became frightened that the bridge might prove too flimsy for their weight. Despite Hutash's warning, they looked down at the ocean, became unsteady, and tumbled into the waters below. Hutash heard their cries for help and transformed them into dolphins, who were forever to lead joyous lives in the seas.[2]

These few stories give just a glimpse of origin myths involving dolphins, which are ubiquitous across the continents. Before we move from mythology to history, however, I will give just one more story, because it has special dimensions that have long puzzled anthropologists and astronomers. It concerns the Dogon people of sub-Saharan Africa in what is now Mali, whose roots reach back more than two thousand years.

The Dogon's origin myth, like that of the Wanungamulangwa people of Australia, has dolphinlike creatures as their ancestors. They came not from the sea, however, but from Sirius, the Dog Star, which is some 8.6 light-years distant. Two French anthropologists spent time with the Dogon in the 1930s and slowly pieced together their stories. The Dogon's knowledge of Sirius appeared to be astonishingly extensive given their lack of technology, and the story of their origins very complex and difficult to follow. Briefly, though, dolphinlike beings from Sirius, called the Nommo, arrived on Earth in starships, which Dogon drawings show landing on three legs. The Nommo populated the seas and became dolphins, and they created children to live on the land, the Dogon people, who were originally called Ogo. This story is found in the Dogon's oral tradition, as well as in symbols carved into doors, lintels, and masks, and in their paintings.[3]

Mythology is, of course, not truth in the way we normally think of truth; that is, it does not generally report events that actually happened or facts that can be verified. But mythologies reach to a different, deeper kind of truth, one that relies on resonance, not on demonstrable evidence. Mythologies do not account for the origin of people or dolphins in the way that scientific theories do, but mythologies tell us something about who we believe ourselves to be, our values, and our place in the world in relation to all the other creatures of nature. Mythologies are, in a way, an expression of that Delphic counsel: Know Thyself.

Given these few legends I've just related, who can doubt the depth of humankind's positive and interdependent connection with the dolphins? These ancient myths represent our perception of dolphins as minds in the water — intelligent, wise, and compassionate.

Few animals bear such numinousity. What is it about dolphins that prompts this kind of reaction, response, and perception?

❦

After mythology comes history, the putative record of actual events. But distant history can sometimes shade into myth, especially when the record is penned many years after the supposed events, as was the case with "true" stories from ancient Greece. Whatever the stories' veracity, there were indeed many tales told from this era of dolphins coming to the aid of men and boys (females were rare in this arena) or simply joining in friendship with humans (usually boys), all of which were in the spirit of complete selflessness on the dolphins' part. Such events were often celebrated by the production of bronze statues of boys and men riding on the backs of dolphins and the minting of coins bearing the images of dolphins. At one point, more than forty cities had coins of this ilk; images of dolphins on coins were as familiar to the Greeks as lions and eagles are to us today.

One of the best known of such stories, not least because it is mentioned in the first act of Shakespeare's *Twelfth Night,* is the rescue of Arion of Methymna, poet and musician of great renown. He spent time in Periander's court, traveled the Greek colonies, and competed in national games (which included music as well as athletics in those days), and he was usually victorious. Through his great talents he amassed significant wealth; he gathered it up and boarded a ship at Taras, which is in the heel of Italy, that was headed for Corinth. Like the unlucky Dionysus before him, Arion realized too late that the crew he had entrusted with his life and his wealth were in fact pirates. Arion begged in vain for his life, offering to give up his money. The pirates would have none of it, because they knew that once Arion arrived safely ashore he would report them to the king. Arion was given two options: he could kill himself onboard and be given a burial ashore, or he could jump overboard right then and there. Not much of a choice.

Arion opted for the latter but asked that he be allowed to sing one last song before he jumped. The pirates agreed, thrilled by the prospect of hearing the world's most famous singer perform before he acquiesced to their nefarious will. Arion put on his full performance costume, took up his lyre, and sang "Orthian," a high-pitched song to the gods. Arion then did as he'd been bidden and jumped into the sea, and the ship sailed on to Corinth, all of Arion's wealth in the hands of the pirates. As the story goes, a dolphin suddenly appeared and approached Arion, took him on its back, and delivered him safely to Tainaron, at the southernmost tip of Greece.

Still wearing his performance costume, Arion made his way to Corinth over the land and went to Periander's court. The king wasn't sure whether to believe Arion's fanciful tale, so he had him locked up and waited for the crew. When they arrived, the king had them brought before him and asked them the whereabouts of Arion. He was safe, they claimed; they had set him down at Taras. At this point Arion stepped out of the shadows and confronted the astonished crew. They were forced to admit what they had done.

Now, these events were said to have taken place around 600 B.C.E., but were not put into writing until Herodotus did so some two hundred years later, based on accounts he heard from people of Corinth and Lesbos, Arion's home. Was it fact or myth? Whatever the truth, there was for five hundred years in the temple at Tainaron a small bronze figure of a man riding a dolphin, put there by Arion himself shortly after his reputed adventure.[4]

There were other such stories from ancient Greece of dolphins selflessly rescuing humans, but more common were tales of dolphins befriending boys, sometimes with tragic outcomes. The first important one occurred around 200 C.E. and involved a boy named Dionysios who lived near Iasos. After school during the summer months, Dionysios and his friends went to the nearby beach and swam in the sea. One day, according to one version of the events, a dolphin approached Dionysios, who, though initially cautious, soon lost his fear of the animal. Before long, the dolphin turned up every day after school to meet Dionysios and take him on his back far out into the sea, returning him safely to the beach each time. The dolphin was said to have fallen passionately for the boy, and the relationship attracted great interest from the townsfolk. At first, crowds gathered to watch this amazing sight, but it soon became commonplace.

On one unfortunate day, the dolphin slid too far up on the sand while returning his friend to safety. Dionysios was unable to get the dolphin back into the water, and the dolphin died. The boy was heartbroken. The story eventually reached the ears of Alexander the Great, and he took it as proof that the sea god Poseidon had taken a special interest in Dionysios, and he appointed the boy to be high priest of Poseidon at the temple in Babylon.[5] Dionysios's story prompted other such claims, as is often the case with unusual events, one of which again came from the people of Iasos, though it appeared much later than the first.

The boy's name was Hermias, and, like Dionysios before him, he rode on the back of his dolphin far out to sea and was returned safely each time. One day, however, a storm came up suddenly, and Hermias was swept off the dolphin's back and drowned. Plutarch described the subsequent events: "The dolphin took the body and threw both it and itself together on land and would not leave until it too had died, thinking it right to share a death for which it imagined it shared a responsibility." In relating this account in his book *Dolphins: The Myth and the Mammal*, Antony Alpers noted, "This is a good example of the great difference there can be between the things that animals do and the meaning that humans will read into them."[6] Alpers did not dispute that the events took place as Plutarch described, simply noted that imputing human motives and emotions to animals was probably going too far. In any case, the people of Iasos commemorated the tragedy by minting coins showing a boy riding a dolphin.

According to Alpers, this next story, the first tale from ancient Rome, is not to be doubted: Two thousand years ago, a young peasant boy lived near Lucrine Lake, a shallow inlet near where the city of Naples now stands. Each day the boy had to walk around the lake to reach his school at Pozzuoli. Living in the lake was a dolphin known locally as Simo, a Greek word meaning "snub-nosed." The boy took to calling, "Simo, Simo," at the water's edge, and, it is said, the dolphin came and ate bread from his hands. Before long, Simo began to ferry the boy across the lake, taking him to school in the morning and then back home in the afternoon. This went on for several years, until one day the boy fell ill and died. According to Pliny, who recorded the events, every morning afterward, the dolphin arrived at the place where he had always met the boy, apparently still looking for him. The dolphin had "a sorrowful air and manifesting every sign of deep affection, until at last, a thing of which no one felt the slightest doubt, it died purely of sorrow and regret."[7]

Most of what is in this story is probably true, except, as Alpers pointed out, that the dolphin ate bread — very unlikely, although certainly possible, as dolphins will sometimes ingest unusual foods or items. And whether it died of "sorrow and regret" is a matter of interpretation and anthropomorphism. The point here is not to focus on the occasional and probably inevitable anthropomorphism of the storytellers. Instead, this tale and the others I've related reveal a time when close relationships between human and dolphin were not uncommon. The two historical accounts I mention here are just the first of many recorded in ancient Greece and Rome. There were numerous others, and we can even make a connection with similar narratives in modern times.

I'll make that connection with events recorded some two thousand years ago at Hippo, a Roman colonial town on the north coast of Africa, not far from modern-day Tunis, and similar events witnessed just fifty years ago at Opononi, a small town on the north coast of New Zealand.

Pliny the Elder wrote the story of a boy and a dolphin at Hippo, and he said in a letter he wrote to his poet friend Caninius that the tale "is true, though it has all the qualities of a fable."[8] The young boys of the town loved to play in the waters, and one game they especially liked was seeing who could be carried farthest out to sea. One boy, bolder than the rest, was far out to sea one day when, to his consternation, a dolphin approached him. The dolphin at first swam around and under him, and then dived, rose under the boy, took him on its back, and swam much farther out. The boy was, quite properly, terrified. But then the dolphin turned around and took the boy safely back to the beach, where the other boys were in awe of what had just happened. Did their friend have supernatural powers? In the days that followed, the dolphin reappeared in the midst of the boys, but they were too timid to get too close. Eventually, the original dolphin rider mounted the dolphin again, and it repeated its previous feat. The dramatic spectacle of the boy riding a wild beast of the sea attracted large crowds of spectators. Unfortunately, managing the crowds of strangers proved to be a financial burden on the town's budget, and sadly the elders secretly decided to do away with the dolphin.

At Opononi in New Zealand, two thousand years later, a young female dolphin, who came to be known locally as Opo, cavorted with young girls and boys in the sea, just like the dolphin at Hippo. Opo seemed to like to be touched and sought out the gentler youngsters for special attention. Jill Baker (a girl at last!) was Opo's favorite, and she would always leave the company of the other children when she entered the water. Although Opo didn't engage in dramatic feats of swimming far out to sea, she did allow Jill, and a few others, to ride her. Two mammal species, separated by ninety-five million years of evolutionary history, playing together, enjoying an extraordinary bond of great simplicity, a rapport that stretches across the ages.

In his survey of the history of dolphins, the eminent anthropologist Ashley Montagu cited the story of Opo and other such contemporary examples, and said, "The so-called myths of the ancients were based on solid facts of observation and not, as has hitherto been supposed, on the imaginings of mythmakers."[9] Yes, storytellers often fall to improperly anthropomorphizing. And, yes, some storytellers no doubt embellish their tales. What storytellers don't? But at its core, the connection between humans and dolphins is undeniable and reaches back thousands of years.*

∞

In the early Christian church, images of dolphins represented positive values such as salvation, teachings about grace, the beauty of the human soul. But before many centuries passed, the reverence and respect for dolphins as sacred beasts that was so prominent in ancient Greece and Rome began to fade away. It's not that reverence for the dolphins had been universal in those ancient times. Dolphins were fair game for the hunt in a few places, which was what led Oppian to condemn it in such forceful terms, but among the leaders of civilization and culture, the bond was indeed powerful. However, starting in the second half of the first millennium and into the first half of the second millennium, all that changed. Stories such as the ones you've just read began to wane. As the force of human activity moved to the beat of dominion over the Earth, as natural resources were seen more and more as ours to exploit, rather than protect, dolphins moved from being sacred to being mundane, just another resource to be exploited for our material benefit.

In the early nineteenth century Frédéric Cuvier, younger brother of the great French zoologist Georges Cuvier, noted the dramatic slide in respect dolphins had suffered from ancient to modern times. Dolphins went from being viewed as a "gentle, good-natured

---

* It was not just in such tales that civilized nations of the past manifested their connection with dolphins. In Greece, the courts of law were based in temples dedicated to the dolphin god. And in France, the title Dauphin (the dolphin) was bestowed on the heir to the French throne from the fourteenth to the eighteenth centuries.

and intelligent animal, most responsive to benevolent treatment" to being dismissed as "merely a voracious carnivore, whose ends are solely those of feeding, resting and reproducing, and whose instincts serve no purpose other than the satisfaction of those needs." It is much easier to slaughter animals if you think of them as voracious carnivores rather than gentle, good-natured, and intelligent creatures.

In his 1973 book *The Cosmic Connection* Carl Sagan pondered what our unrestrained slaughter of dolphins and whales told us about ourselves. Noting that there was emerging evidence that dolphins and whales were far more intelligent than most people had thought possible, he said: "They have acted benignly and in many cases affectionately toward us. We have systematically slaughtered them. Little reverence for life is evident in the whaling industry — underscoring a deep human failing." Know Thyself.

A shift in attitude toward dolphins and whales was afoot as Carl was writing those dark words, a gradual stirring of an ancient but long-dormant worldview. As modern scientists began to uncover and document the remarkable abilities of the dolphin mind, the nonscientific public rediscovered the visceral connection with dolphins and whales that the people of ancient Greece and Rome had seen as part of the natural order of things.

Stories of dolphins saving shipwrecked sailors and keeping sharks at bay when swimmers were in trouble once again began to rise in our consciousness. In the early 1970s, the eerily beautiful songs of the humpback whales struck a primordial chord in all but the most hardened listener. Whale-watching tours and swim-with-dolphins programs were in the nascent stages of what has become a multibillion-dollar business. I know from my own experience the profound feeling of being with a "presence" when I am with dolphins. It is almost impossible to put into words. But I think I know what it means:

Know Thyself, each and every one of us. Know Thyself as a

species with privileges and responsibilities on this Earth, responsibilities to recognize and honor the inherent value of other species.

⸙

Today, tragically, dolphins and whales are being brutally slaughtered and driven toward extinction by modern and otherwise civilized humans. Despite a brief moratorium on whaling in the mid-1980s, today whaling is still practiced by many countries and territories, including Canada, the commonwealth of Dominica, the Faroe Islands, Greenland, Grenada, Indonesia, Japan, Norway, the Philippines, Russia, St. Lucia, St. Vincent, the Grenadines, and the United States. *Moby-Dick* is an American classic about the brutal practice that was discontinued, though there is one exception: each year, nine indigenous Alaskan communities are permitted to hunt a total of fifty bowhead whales. I'm sharing the scientific and personal experiences I've had with dolphins — these remarkable minds in the water — in the hope that you will become as convinced as I am that they deserve global protection and respect.

2

FIRST INSIGHTS

SUMMER 1977.

As I lay in the darkness under oppressively humid tropical heat, I could hear the soft murmur of crickets in the nearby grove of torchwoods; the trees' sweet aroma hung in the air. My simple thatched cabin was dark as pitch; the only light came from a few stars shining through the wooden louvers of the window that overlooked the lagoon just outside. Every few minutes, the constant, soft ratcheting of the crickets was interrupted by another sound: *chuff*. It was way past midnight, and sleep was nowhere near. I'm a night owl anyway, so I'm used to late hours. But this night it wasn't my nocturnal habit that kept me up. *Chuff*. It was eager anticipation.

Earlier that day I had arrived in Little Torch Key, about twenty-five miles from Key West, Florida, to conduct my first study of dolphins. There were two of them, a male and a female, both bottlenose dolphins. They were to be my companions and mentors of a sort for the next month. During that first day I sat by the edge of the lagoon, quietly observing their behavior, and that night I could hear their breathing — *chuff* — as from time to time they broke the water's surface and exhaled and inhaled through the blowholes on the tops of their heads.

My recently formulated life's goal was not modest: I wanted to understand the dolphin mind and learn how these highly social animals communicated. What little was known about these realms at this point came principally from the work of John Lilly, who'd pioneered research in dolphin communication and intelligence. He had initiated his investigations more than two decades earlier and used a combination of electrophysiology, acoustic analysis, and training techniques to study dolphin intelligence and the potential for communication with other species.

In 1960 Lilly speculated that in the near future, the human species would establish communication with another intelligence, "non-human, alien, possibly extraterrestrial, more probably marine; but definitely highly intelligent, perhaps intellectual."[1]

He envisioned humans establishing an interspecies dialogue with dolphins. In what may seem like a sci-fi scenario, Lilly acquired a house by the sea on St. Thomas in the Virgin Islands, flooded its lower floors with seawater, and transformed it into a live-in laboratory where he and his assistant Margaret Howe attempted to teach dolphins to speak English. Lilly housed his dolphins under what would now be considered inhumane and unacceptable conditions — in small, shallow pools. I literally cringe every time I see images of those one-and-a-half- to four-and-a-half-feet-deep pools and the coffin-size Plexiglas testing tanks. But these were the 1960s, and scientific consciousness of what constitutes proper husbandry for dolphins was in its infancy. Lilly speculated that the large and complex-brained dolphin, known for its proclivity for vocal imitation, would, like a human child, be able to learn English if provided with the correct social conditions. To test this, he conducted an experiment during which Margaret Howe lived with a young male dolphin, Peter, for several weeks. Of course, they did not share a level playing field of social interactions and exchanges. Nor was she rearing Peter, as had been attempted previously with chimpanzees to see if they could learn language if brought up in similar conditions as a human child. Instead, she held a fish bucket and trained the dolphin to imitate the number of syllables, or "sonic bursts," that she produced, rewarding him with fish when he got it right. The dolphin was able to master the task, and the results were published in the *Journal of the Acoustical Society of America* in 1968 under the title "Reprogramming of the Sonic Output of the Dolphin: Sonic Burst Count Matching."

In any case, Lilly was the first person in modern times to recognize that dolphins have large, complex brains, that they are highly intelligent, and that they are adept vocal mimics. He was the lone pioneer in this field for many years, and he deserves credit for sparking scientific and public interest in dolphins, their brains, and their intelligence and communication abilities in his early writing, from 1954 to 1968. His suggestion that humans could no longer claim to be the only superintelligent beings on the planet proved to be prophetic. He also professed that dolphins' ability to mimic sounds combined with their intelligence would enable them to learn and use English words. This idea was so fantastic to me when I read it in 1977 that I rushed out and bought a record that Lilly had made a few years earlier, *Sounds and the Ultra-Sounds of the Bottle-Nose Dolphin*. I still can bring to mind Margaret Howe's rich Southern accent as she said to the dolphin, "One, two, three, foe-er," the dolphin responding with four bursts of sound in the same rhythmic pattern. However, I soon became keenly skeptical of the idea that this line of work would ever go anywhere. Indeed, Lilly, who died in 1986, never achieved his dream of having a conversation in English with dolphins.

As unorthodox as his approach was, Lilly was responsible for establishing and stimulating research in the science of dolphin cognition, and through his popular writing he ignited the public's interest in dolphins and their amazing abilities. In his own visionary and eccentric way, he opened up the real possibility that somehow we humans might be able to communicate with a species very different from us, and vice versa.

I wanted to explore that possibility. How? I wasn't quite certain yet. I was aware of the groundbreaking attempts at the time to teach language-like codes to species other than our own. For instance, Allen and Beatrix Gardner and their graduate student Roger Fouts taught a young female chimpanzee, Washoe, to communicate using a modified version of American Sign Language; David Premack at the University of Pennsylvania taught the chimpanzee Sarah to use a code of visual symbols (Premack's theory of mind, the ability to infer the intentions, beliefs, and desires of other individuals, has been highly influential); Irene Pepperberg worked with Alex, an African Grey parrot, whose burgeoning verbal abilities gave her a window into his mind. Yet I already had an inkling that this realm of research was viewed by some as less than scientific. (The antagonistic undercurrents regarding some of these studies exploded into public view with extraordinary force and animosity just three years later, in May 1980, at a now famous conference at the New York Academy of Sciences.)

I was determined to pursue a rigorous line of investigation in my own work, whatever I did, so much so that colleagues have sometimes teased me as I doggedly gather one more piece of evidence to support an already pretty secure conclusion. At the same time, I knew in every fiber of my being that communication between two individuals is a social process, facilitated by familiarity and trust between them. After just a few days of observing the two dolphins in that lagoon on Little Torch Key, I began to feel that familiarity and trust, especially as I recognized that they were also observing me. I sensed a familiarity in our interactions, a pattern of behavior that seemed easily recognizable. I had become entranced with the writings of the British anthropologist, social scientist, and thinker Gregory Bateson, a man who, among his other accomplishments, spent some time observing the social interactions of dolphins at Sea Life Park Hawaii. A phrase from his last book captured what I was experiencing in these first encounters with dolphins and would continue to experience throughout my work with them: "What is the pattern which connects all living creatures?"[2] The pattern that connects; the recognition of familiarity.

My agenda, then, as I embarked on my journey was to learn everything that was currently known about dolphin communication and behavior, and then to go beyond those frontiers into the unknown, into the realm of dolphin mind. I wanted to explore the far reaches of their minds, to dive into those unknown waters and find out what they can do and what they know. A few years earlier, Thomas Nagel, a professor of philosophy and law at New York University, had published what would become a classic paper in the realm of animal behavior and cognition and philosophy. It was titled "What Is It Like to Be a Bat?" He explored the notion that perhaps there were experiences beyond human understanding, intellectually and viscerally. We can try to imagine, he argued, what it would be like to be blind and equipped mainly with exquisite sonar (echolocation) for navigation and detecting insect prey; we can try to imagine what it would be like to eat bugs night after night and hang upside down in a cave during the daylight hours; and we can try to imagine what it would be like to flap our arms and fly with superb agility. But this exercise suggests only what it would be like for a *human* to be a bat, not necessarily what it is like for a *bat* to be a bat. Part of my long-term goal was to achieve the apparently impossible: to know what it is like for a *dolphin* to be a dolphin.

When I arrived in Little Torch Key that summer of 1977, my Honda station wagon was loaded with heavy-duty equipment for recording underwater sounds: a big J-9 hydrophone, a huge eight-track reel-to-reel recorder, and forty tapes. I was a doctoral student in the Speech and Communications Department of Temple University and I had received a two-thousand-dollar biomedical research grant for women in science from the National Institutes of Health to conduct observations and record the vocalizations of two semi-feral bottlenose dolphins. Such a modest sum wasn't going to provide me with the kind of equipment I needed for recording dolphin whistles, however. I rustled up the equipment by way of the military's technology transfer program; I called dozens of military bases around the country until I located everything. I had my own form of sonar to zero in on the equipment; I already had experience in recording and analyzing human speech. My goal during my month in Florida was to further develop expertise in recording and analyzing dolphin whistles. But during that first week, all the expensive (and now completely obsolete) equipment sat in my little thatched hut, unused. I had thought long and logically about how to carry out this mini research program, how to document the patterns of dolphin behavior and record the whistles that accompany them. But the approach I finally adopted really came from my gut, my intuition.

Scientific methods vary. Each one typically demands that a researcher adopt a well-established, regimented approach to collecting data. In the field of animal behavior, for instance, a researcher may observe an individual animal or a group of animals at set intervals (say, every thirty seconds) and then choose items from a predetermined catalogue of behaviors, called an ethogram, that describe those behaviors; this sequence of behavioral snapshots becomes the raw data for analysis. Another option is to videotape the ongoing and interactive behavior of animals and then analyze the behavior back at the lab. I didn't use either approach. Instead, during the first week of the project, I spent a lot of quiet time sitting by the edge of the lagoon, simply observing the two dolphins, mentally noting what they did when they swam separately; how they interacted together, as they so often did; and how they interacted with me, as they were so obviously keen to do. I wanted to *absorb* something of their overall patterns of behavior, not tabulate them in the conventional manner.

These creatures were so strange to me in so many ways — in

their physical form, how they moved, and what was going on in their minds — that I was, figuratively speaking, blind and deaf to their behavior. It was like working with aliens. By initially simply observing them, *being* with them, rather than *studying* them, I believed I would be able to begin to close that gulf between us so that I could start to see and hear their alien world a little bit. Only then, I reasoned, would I be in a position to act the conventional scientist with them, aided by technology and routine observational regimens. I hadn't had any role models in this new (to me) realm of animal behavior. I was guided, in part, on behavioral research methods by the writings of Princeton ethologist Jeanne Altmann. But largely, I was on my own.* It was a humble start. In the years to come I would greatly enjoy rich collaborations with colleagues and students.

It had been John Lilly who had guided my path to Little Torch Key that summer. Months before, I had tracked down Lilly's phone number in California and called him. "Dr. Lilly," I said. "You don't know me. I am Diana Reiss and I am going to do research on dolphin communication. I need to talk to you." Lilly was very patient and encouraging in what was quite a long conversation. I was thrilled and more determined than ever to follow my impulse.

We had several phone conversations after that first call, and he was charming and more helpful than I deserved. I met him in person shortly after this dialogue began, at a public lecture in Manhattan, and I told him I was looking for guidance as to how to get started with working with dolphins. "You should contact Betty Brothers in Florida," he suggested. "She has two dolphins in a lagoon next

---

* These days I tell my students when they begin their first observations, "Don't take anything. Don't take pencils and papers. I just want you to go out and I want you to sit there for a week. And I want you to just watch them. There's no pressure. Just watch and see what you're seeing. And then write up ideas that you have of what you've seen, what kind of behaviors you saw." I don't want to give them an ethogram; it would filter the information too much. I'm more interested in what they see for themselves.

to her house. She's had them for years. I'm sure she'd be delighted to have you go and do your work there. I'll arrange it." Naturally, I was thrilled. He added, "Betty is a remarkable woman. You'll love her!"

John was right. She was, and I did. Betty had lived in the Keys since 1952 and loved the proximity of wild dolphins to her ocean-front house. She loved to watch them as they swam past her property each day. In the early 1960s, the dolphins stopped visiting, and Betty and her husband decided to acquire two of their own, which they purchased from a dolphin training school.* The first, which they obtained in 1962, was a female they named Dal. Dal was joined three years later by Suwa, a young male. The two bottlenose dolphins spent their days and nights in the large lagoon by Betty's house, essentially captive. Then, in 1966, Betty removed the barrier between the lagoon and the ocean, allowing Dal and Suwa to go where they wanted to. Aside from a period early on, when they disappeared in the ocean for six days, Dal and Suwa were in the ocean during the day and in the lagoon at night. Eventually, the two dolphins spent more and more time only in the lagoon, apparently content to trade the complete freedom of the ocean for the certainty of two fish meals each day.

I was excited to be at the lagoon. When I arrived, Betty, who personified the phrase *feisty woman,* welcomed me as a daughter. Like me, she was an animal lover. She had two dogs running around the place, so I was in heaven. It was so hot in the Keys that Betty routinely put ice cubes in the dogs' water bowls. Betty was in the real estate business, and her home and office compound included a few small rental cabins on the edges of the lagoon. She asked if I would mind helping her out by feeding the dolphins during the day, thus freeing her for her real estate activities. Would I *mind?* I

---

* This was before the Marine Mammal Protection Act of 1972, which prohibited the capture of wild dolphins for private individuals.

was delighted to have the chance to become intimate with Dal and Suwa. Betty's father, Carl, also lived at the house, and he took me under his wing and taught me how to cut the fish the way the dolphins liked it, which included removing the spiny dorsal fins from the large Spanish mackerel they were fed.

By an interesting coincidence, Betty's lagoon was just a short distance from where, a decade before I arrived, the *Flipper* television series had been filmed. As it happened, Betty had acquired her dolphins through the same people that captured the dolphins for the show. Visiting Betty was like having Flipper in my backyard! How ironic, given my lack of regard for the *Flipper* series yet my passion for studying the real thing.

During that first week of quiet observation and the subsequent weeks of more directed study, I noticed how very social the dolphins were, not just with each other but also with me. I quickly realized (actually, it was an uncanny *feeling*) that in the same way I was observing them, they were observing me. They made extraordinary direct eye contact. I got a powerful sense of a very real presence there.

It was impossible to ignore. They seemed to deliberately engage me in games, such as the seaweed game, in which they'd take mouthfuls of seaweed, come to the side of the lagoon where I was sitting, and flip the seaweed to me. It was obvious what I was expected to do: flip it right back. I was essentially imitating what they were doing. This back and forth, back and forth might go on for several minutes. I imitated other behaviors too. For instance, they would occasionally lie on their sides and look up at me, and I would then do the same thing, lying on my side and looking down at them, trying to match their posture and body orientation. What might seem like trivial behavior patterns have elements that are profound. The initiation and imitation was establishing a basic level of communication, one species to another: I see you. I acknowledge you. I'm doing it back. From the beginning I was guided by a simple rule:

All animals communicate, and we can communicate with other animals. It is not a matter of *whether* we can communicate; it is a question of *how*. And here, by the lagoon, an instance of "how" was emerging.

❧

Dolphins* have a certain ineffable presence about them, which is partly what draws people to aquariums and drives them to seek swim-with-dolphins experiences. Despite this visceral connection, people recognize how different humans are from dolphins in physical form and, of course, in aquatic abilities. But most people are unaware of how very different dolphins are from humans in mundane realms, such as breathing and sleeping and the perception of their environment.

I'll start with the differences in how dolphins and humans breathe and sleep, which, odd as it might seem, are actually closely linked. A dolphin breathes through a blowhole that is positioned on the top of the head, an important aspect of dolphins' evolution from land animal to marine mammal. The nostrils of dolphins' ancestors slowly "migrated" from the front to the top of the head. Nostrils atop one's head makes breathing easier during swimming or resting at the surface. Having your nostrils on top of your head might seem bizarre, but that is not the most important difference between us and them. The most important difference is the *manner* in which dolphins breathe. For humans, and for most terrestrial mammals, breathing is an involuntary process. The rate and timing of human breathing is governed by receptors in the brain that monitor the level of carbon dioxide in the blood and respond appropriately. Humans don't have to think about it. But for an air-breathing mammal who spends most of its time underwater, exactly *when* it breathes

---

* When speaking of dolphins, people generally mean the bottlenose dolphin, because that is the species that's been the most popularized. In fact, there are more than thirty species of dolphin.

can be a matter of life or death. When receptors in a dolphin's brain indicate it needs to breathe, the dolphin must first swim to the surface. In other words, for dolphins, breathing is a voluntary process, not an involuntary one.

A human wouldn't make a very good dolphin because he or she would have to surface half a dozen times each minute to breathe, which would be very inconvenient for a creature that forages and conducts its social life underwater. Dolphins, however, are superbly adapted to life under the waves. Under normal circumstances, they breathe two to four times a minute, although they can hold their breath as long as fifteen minutes; they exhale (just as they reach the surface) and inhale in a fraction of a second; they exchange up to 90 percent of the air in their lungs, compared with our much more limited 10 percent; and they extract much more oxygen from each breath than we do. The anatomy and physiology of their breathing is much more efficient than ours.

So when we ask what it's like to be a dolphin, part of the answer is that they have to think about breathing. Also, they sleep very differently than we do. All mammals sleep, and dolphins are no exception. When humans go to sleep, we become unconscious, and our physiological systems run on autopilot. Dolphins have no such autopilot; at least, not in the realm of sleeping. To keep the dolphin from drowning, there has to be some part of the dolphin brain that remains awake. And yet, they also need to sleep. How do they do it? Dolphin brains, like ours, are divided into two hemispheres. When we humans sleep, both hemispheres go into unconscious mode. Through evolution, dolphins have arrived at another strategy: one hemisphere sleeps or rests while the other maintains breathing duties, motor behaviors, and possibly also some vigilance against predators. When dolphins are in sleep mode they often lie motionlessly at the water surface or at the bottom of a pool, or they might swim slowly, sometimes with one eye closed, surfacing from time to time to breathe. Dolphins appear to sleep or rest many times throughout the day and night. We don't know if they dream.

Another big difference is in how they perceive the world. Dolphins, like us, face complex environmental and social challenges, and they use their senses and perceptions to survive and thrive in their dynamic environment. Dolphins use a rich constellation of multisensory signals to communicate with one another, including acoustic, visual, tactile, and possibly chemical (taste) signals. In the course of the dolphins' adaptation to a totally marine existence, they have evolved specialized sensory systems that enable them to perceive their world even when vision is limited and vast distances separate them. Dolphins have the basic senses of hearing, vision, touch, and taste, but they have lost their olfactory receptors and the olfactory areas in the brain—thus they are unable to smell. Terrestrial mammals use scent trails and territorial scent marks, but these do not exist in the aquatic world, and the unneeded sense was eventually lost.

The visual systems of animals are adapted to each species' environmental requirements. As primates, humans have excellent vision (not as good as some birds', of course), and it is our primary sensory guide in the world. Humans, like most other primates, have trichromatic vision; this means the eye has three types of color cones, enabling an individual to perceive a myriad of colors. The dolphins' sensory world is highly acoustic, so many people believed that they had inferior visual abilities. But in fact, dolphins have excellent vision. Although they are monochromats—they have only one type of color cone, and thus are colorblind—they are very sensitive to light in the blue region, which is the color of their aquatic world. The dolphin eye has many more rod cells than the human eye, making it very sensitive to low-light conditions, an important adaptation for life in the sea. Underwater, a human can't see very well without a facemask, but the lens of the dolphin eye is highly specialized and affords the dolphin vision that's just as good in the air as it is in the water.

Humans have at best a mediocre sense of hearing when compared with that of many other animals, but our hearing really pales in comparison to the acoustic processing abilities of dolphins. Their marine world is filled with sound; sounds travel farther and about 4.3 times faster in water than in air. Our world is made up of visual images; the dolphins' world is one of acoustic images, as well as images.

For creatures whose perceptual world is so dominated by sound, the absence of any obvious external ears might seem odd. The ears have been lost over evolutionary time, sacrificed in the radial streamlining of their bodies in their adaptation to a superbly efficient hydrodynamic physical form. (Dolphins are one of the swiftest creatures in the oceans.) The external ears have vanished, but the inner ears are present and again exquisitely adapted to the species' needs. For instance, dolphins can locate the source of a sound underwater, which humans find almost impossible, and their range of hearing is much greater than ours — it stretches from 150 to 150,000 Hz or higher, compared with humans' 20 Hz (a little lower than the first key on a piano) to 20,000 Hz (about two octaves higher than the last key on a piano). It is their high-frequency range that also sets dolphins apart from humans.

In the course of evolving into fully aquatic mammals, dolphins have developed an exquisite and highly sophisticated biosonar system known as echolocation that allows them to navigate in a world without particularly good vision. It's a biological version of man-made radar and sonar, but much more powerful. (Dolphin echolocation is the envy of the military.) Dolphins have air sacs beneath their blowholes, and a dolphin echolocates by moving air between these air sacs at an extremely fast rate; this produces rapid sequences of high-frequency clicks, each of which lasts less than a thousandth of a second, that are directed forward in a narrow beam (the beam is shaped partly by the contours of the dolphin's bony skull and partly by a fatty "melon" that acts as an acoustic lens to focus sound). When the sound waves hit a solid object, echoes bounce off the object, travel back to the dolphin, are collected through the bones of the lower jaw (and a few other areas of the body), and are transmitted to the inner ears.

The numbers here are impressive. Dolphins can produce from as few as eight clicks a second, which sounds like a creaking door, to as many as two thousand clicks a second, which sounds like a high-pitched buzz. Individual clicks are wide-band sounds, composed of a rich mixture of low- and high-frequency wavelengths. The low-frequency clicks are longer wavelengths; they travel farther and give the dolphin a general overview of an object. The high-frequency clicks are shorter wavelengths; they don't travel as far but they provide more details. One might expect the returning echoes to produce a confusing cacophony impossible to interpret. But in fact, the dolphin receives a clear acoustic image of its surroundings. Dolphins can almost literally see with sound. And as I mentioned above, the exquisite degree of image discrimination has navy technicians salivating. To give just one example: at a distance of thirty feet, a dolphin can detect a few tenths of a millimeter's difference (about the thickness of a fingernail) in the density of the walls of two metal cylinders.

In addition to being highly sensitive to sound and visual images, dolphins are highly sensitive to touch. Touch plays an important role in human social interactions and relationships, and the same seems to be true in dolphin societies. Dolphin social behavior frequently involves tactile interactions such as pectoral-fin rubbing, pectoral-to-pectoral fin contact while swimming (it looks like handholding!), and body rubbing. It's been reported that the level of tactile sensitivity on some areas of the dolphin body is comparable to the sensitivity of our fingertips and lips.

Imagine being able to control your breathing and hold your breath for astonishingly long periods; imagine sleeping half a brain at a time; imagine acoustically "seeing" finely detailed images of distant objects in complete darkness with your biological sonar. It has been suggested, although not verified, that dolphins have the ability to image the internal bodies of others, so imagine having x-ray vision and being able to scan the internal bodies of your family members and friends; you'd know about all their pregnancies, injuries, and illnesses. Imagine, too, that your environment is the ocean and that you are the king or queen of speed. That — in part, at least, and from a limited viewpoint — is what it might be like for a human to be a dolphin. But how can anyone really know what it's like for a dolphin to be a dolphin? My goal was to get a glimpse of that by opening a window into the dolphin mind.

When my month at Little Torch Key was over, I hadn't discovered any significant insights into the cognitive or communicative abilities of dolphins. But I had accomplished my research goals for the project: I'd learned the technical aspects of recording dolphin vocalizations using specialized underwater equipment, and I had a car full of large reels of tape with recordings of those vocalizations to take back to the university for analysis. I had put the equipment to good use, recording their different sounds and testing to see if Dal or Suwa used specific types of vocalizations when fed or when presented with a particular object, such as a ball, a ring, or a seashell. I had not collected publishable observations during that month, yet the experience was an invaluable foundation for work that, I hoped, would eventually merit publication in scientific journals. The experience also left me with more questions than I'd started with, and for that I was thrilled. For a scientist, knowing the questions to ask is as important as finding the answers (and sometimes more important).

On a personal level, Dal and Suwa provided my first encounter, and I experienced a sense of their intelligence. It wasn't just the experience of being held in an uncanny gaze; it was a sense of a familiar presence in these creatures, a manifestation of the pattern that connects. But I was unsure of my next steps. I wished I had the seal of Solomon, a signet ring that was said to give King Solomon the power to communicate with other animals. Konrad Lorenz, the famous German ethologist, had claimed to accomplish this feat, not by using Solomon's ring but by using his own powers of observation of the animals around him. If you pay close attention to their interactions with one another and with you, animals will often reveal their secrets.

Toward the end of the month, I had taken to amusing myself in the evenings by playing my flute by the lagoon. It was simply for relaxation, although it was romantic in the purest sense, there by the water's edge with the western sky still brushed crimson from the sunset. To my surprise, I began to get the impression that Dal and Suwa were paying attention to my playing. I discovered by simple experimentation that if I picked up the tempo, they swam faster. And if I played slowly, they became somewhat languorous.* I remember thinking on one such evening, *Wow, here I am studying dolphins, looking out at the open ocean and playing my flute, which these creatures are responding to from their inner nature. I am the luckiest person in the world!* I was hooked.

❧

It was not long after my return from Florida that I realized the path to take. Lacking the ring of wise old Solomon but agreeing with Lorenz that the powers of observing and listening to animals were critical, I wanted to use a combination of observational and experimental approaches: first, observing and recording the behavior and vocalizations of the dolphins within their own social interactions, and, second, finding a way for the dolphins to interact with a symbolic system that would allow them to have at least some rudimentary choice and control over their environment. I sought to

---

* Three decades later, animals' attunement to human music became a subject of serious scientific study.

develop an underwater keyboard for dolphins that would provide them with some ability to communicate with us via a shared code of symbols. My beginning premise was that dolphins were an intelligent and highly social species who were already communicating among themselves. About what, I had no idea. I have an abiding philosophical and pragmatic fear that we may be unable to recognize other forms of intelligence in a species that is so unlike us and that is coming from such an alien environment. But it had become increasingly clear to me that a shared keyboard might be a means to allow the dolphins to "show us their best stuff."

The Animal Acoustics Laboratory of France's National Center for Zoological Research is located in a chateau near the small town of Jouy-en-Josas, some two and a half miles southeast of Versailles and about twelve miles southwest of the center of Paris. This charming village is best known for its famous fabric design, called toile de Jouy, a repeating pattern of a pastoral scene in a single color, usually set against a white or off-white background. Much of the countryside surrounding Jouy-en-Josas is heavily wooded, but open parkland and rolling hills here and there are redolent of a grandeur appropriate to a region of such fabled history. The French chateau itself looked like something out of Grimms' fairy tales, with steep roofs and elegant long windows beyond intricate laceworks of black iron gates. The lab, by contrast, was like any science lab, with benches, bookcases, and electronic gadgetry.

René-Guy Busnel was the director of the lab at the time, and he was the reason I was there. He had done much work on dolphin acoustics, had worked with insect and bird communication too, and he'd edited the volume *Acoustic Behavior of Animals*, published in 1966, making him the world's authority in the realm of animal bioacoustics. But that wasn't all. His monograph *Whistled Languages*, published in 1976, is still the most important work about this little-known realm of human communication.

In 1976 I'd known that John Lilly was the person to speak to about my nascent interest in dolphins, and I now recognized that I had to contact Professor Busnel. I wrote a letter, and he later invited me to visit him at his lab.

I explained that my goal was to decode dolphin communication and learn more about the minds of dolphin and that I planned to follow two separate paths. The first, which had already had a little traffic, was to record dolphin whistles as a step in understanding the information they contained. I recognized that this was extremely challenging because we lacked a Rosetta stone to help us decode their calls. I therefore needed a parallel line of investigation that might inform the first. This second path, which at this point was almost virgin territory with dolphins, was to develop an artificial code, probably using a keyboard as Duane Rumbaugh and Sue Savage Rumbaugh were doing with chimpanzees. With an artificial code, I hoped to learn something about how they communicated that I could then employ as a window into their natural means of communication. I also thought it was important to give them more choice and control if they were in captivity. Busnel was interested in these ideas and suggested I do some of my graduate work at the lab. He helped shepherd through a grant from the French government and also enlisted my help with a NATO sonar systems conference on echolocation that he was organizing.

Beginning in September 1978 I took the bus to the lab each day, leaving my small studio apartment on rue Pernety in the Fourteenth Arrondissement of Paris. This neighborhood lies just south of the Tour Montparnasse, an area famous for its ateliers of artists, actors, and philosophers. Through a friend of Professor Busnel and his wife, Marie-Claire, also a renowned scientist in bioacoustics, I had found the inexpensive studio apartment located in a very old five-story house. My room was small but charming. The good news: it had a wood-burning fireplace, two french windows overlooking a small courtyard with a large fig tree, a raised built-in bed with bookshelves on the surrounding walls, a small sink with hot and cold water, and a shower. The bad news: it had no kitchen or bathroom. The sink was near one of the windows, and a three-by-three-foot shower basin was on the floor near the sink. As for a toilet, it was a shared one conveniently located outside on the stairway landing between my floor and the next floor up. It was essentially a hole in the floor, flushed by a brisk pull of a string dangling from the tank above. This type of toilet is commonly referred to as a Turkish toilet by the French and as a French toilet by Americans. (No country wants to claim it as its own.) I was thrilled to have the studio, and it was a bargain, about ninety dollars a month.

The daily bus journey to the lab was a good opportunity to improve my rather limited French. When I arrived at the center, I walked through the gates past the chateau and onto a smaller road that wound up a hill to the green rectangular building that was the acoustics laboratory. My office was a small square room on the second floor with a U-shaped counter-cum-desk that faced a window onto rolling lawns. It was a haven for me. In France, doctoral students were treated with great respect and, unlike their counterparts in the United States, given time to just read and think. The graduate training grant I had been awarded by the French government provided for American students to go to Paris and study in a specialty area. My area was about as specialized as you could get: dolphin communication and human whistled languages.

One vivid memory I have from my walks to the lab was of some very interesting behavior of the crows that lived in the surrounding woods. Busnel had told me to watch out for it. The crows had been seen repeatedly dropping or placing intact chestnuts in the roadway and then waiting in the treetops above until a truck or car ran over the nuts and cracked them open. They would then swoop down and feast on the exposed kernels. I saw the crows engaged in this activity on many occasions. This was at a time when reports were on the rise of birds and other animals using tools, joining us humans in the tool-use domain.

Busnel was the first person to study human whistle languages, beginning in the 1950s. I read some of his papers on the topic before

I met him. Spoken language is one of the most plastic and rapidly evolving aspects of fundamental human behaviors. More than five thousand distinct languages exist around the globe today, probably a rather small representation of what used to exist in the not so distant past. Their variants are enormous and involve components that to the average Germanic- or Romance-language speakers are quite exotic, such as clicks, gutturals, and nasal sounds. But all these elements of language are for the most part products of a natural, unconscious evolution. Not so with whistle languages. Whistle languages, of which there are scores around the world, are deliberate adaptations to particular contexts in which normal spoken language, including shouting, does not work. The most common ecological context is in mountainous regions, where accomplished whistlers can communicate detailed instructions or pass on complex information across a valley or from the top of the mountain to the bottom of the valley, the sounds often traversing as much as three miles.

Busnel's initial interest in whistle languages was as a bioacoustician studying variants of human vocal communication, but his excursions into the acoustics of calls in other animals, especially dolphins, put the whistle languages in a new intellectual context. When I first arrived at the lab in Jouy-en-Josas I was thrilled to find that there were some whistlers from Eastern Europe visiting for a short while. Busnel had filmed these people in their own country and now wanted to record them for spectrographic analysis in the lab. I was rapt. I heard men produce quite lengthy but not obviously complicated whistles that in fact were whole sentences containing a lot of information. (Women from Eastern Europe apparently don't whistle. Why, I'm not sure.) Here were people who, under the constraints of a certain environment, had developed a means of encoding a lot of information in what seemed to be simple whistles. And I was thinking, *This is astonishing. Dolphins were once terrestrial animals, and then they evolved to inhabit a very different medium, the sea, which resulted in all kinds of adaptations, the most obvious being body form. But they, too, communicate with whistles — sounds that travel farther in water than other sounds.*

Now, dolphin communication as compared with that of mountain-dwelling humans is quite distinct, and the vocal adaptation of dolphins was not intentional, of course. How the vocal repertoires of dolphins have developed and how they are related to the communication signals of their ancient terrestrial ancestors remains a mystery. We simply do not have an acoustic record, analogous to a fossil record, of early cetacean signals.

But what I heard in the lab that day was enlightening: simple whistles are effective at long-distance communication and can encode large quantities of information. And here's another difference between human whistlers and dolphins: after a human whistler has dazzled us with an especially complicated riff, we can ask him to translate the message, and he will. The challenge that stretched before me back in the lab in Jouy-en-Josas in 1978 was to discover what dolphins mean when they whistle.

⤳⤳

In Greek mythology, Circe was an enchantress and sorceress who lived with her nymph attendants on the island of Aeaea. Her powers of seduction were legendary, and if it so pleased her she would turn men who offended her into animals. Such was the fate of some crew members of Homer's hero Odysseus (Ulysses) as they journeyed home after the war in Troy. The men, hungry for more than the food and wine that the beautiful Circe was offering after their long journey at sea, got more than they bargained for: they found themselves transformed into swine. Odysseus rescued his men and, in one version of the story, went on to sire many children with Circe. With flaming red hair and magical powers, not to mention the ability to brew intoxicating potions, Circe was the very definition of *enchantress.*

In my world, Circe was a young female dolphin, the first with whom I spent serious research time, the first to give me a glimpse of a dolphin mind beyond anything I had imagined. She was the dolphin that set the intellectual stage for all my later work. She enchanted me right from the beginning, and because I'd been fascinated by Greek gods and goddesses since the sixth grade, I named her Circe. The metaphor goes further for me, because, as I said, the goddess Circe had been able to transform humans into animals, and what I was attempting to do with the dolphin Circe was create a means of communication to connect us and demonstrate there was continuity between us as thinking, cognizant creatures. (Beyond that, in an academic pun, CIRCE was an acronym for Cetacean Intelligence Research and Communication Experiment.)

Port Barcares is a picturesque town on the coast near Perpignan in southwestern France. It was, appropriately, a magical location, with the rugged Pyrenees to the west and the blue waters of the Mediterranean just a stone's throw to the south. When I first arrived, in January 1979, I literally gasped at the beauty I witnessed around me, even though the area was wearing its stark winter coat and was very cold. The small marine zoo there was a mom-and-pop affair, run by Willy Stone, a large, bighearted, white-haired man with a walleye, and his petite, demure Dutch wife, who seemed quite out of place in this remote setting. Monsieur Stone and his wife lived in an incongruously small caravan, the most modest structure on the site. Their teenage daughter, who looked like a young Brigitte Bardot, lived in another caravan. They all cared passionately about the animals, which mitigated my somewhat negative response to the, shall we say, limited facilities that housed four dolphins. Peacocks, black swans, and white swans roamed freely, which gave the place the atmosphere of a French farm, but one had to be vigilant because the black swans were quite aggressive and seemed to enjoy attacking people's feet.

During the previous six months at the animal acoustics lab I had been immersed in communication theory, information coding, and dolphin sound-production mechanisms and vocalizations, and

I had started to master the practicalities of recording and analyzing cetacean vocalizations. But I needed to begin working with dolphins, not just reading about them. The research I was to conduct would form the basis of my doctoral thesis. I had several research goals, and all of them were related to my longtime vision of developing the underwater keyboard system.

I had imagined using three-dimensional white geometric forms on a black background as symbols on the underwater keyboard. Although I was sure the dolphins could use echolocation to discriminate among the different forms underwater, I wanted to make sure that they could also visually discriminate among them. So my first experiment was to investigate whether dolphins could visually discriminate one three-dimensional white geometric form on a black background from another when the forms were presented to them above the water's surface. A second experiment tested whether dolphins could learn conditional discrimination, which, simply put, meant seeing if they could learn to associate a specific visual symbol with a specific toy (say, a ball or ring). The third experiment was more complicated. Initially, I'd determine the dolphin's toy preferences — in a given set of toys, which objects did the dolphin play with most frequently? After this, I would provide the dolphin with a free-choice situation, a rudimentary keyboard of sorts, that would display the three visual symbols that had been associated with the different toys in the previous experiment. The visual symbols would be arranged horizontally in varying positions on the keyboard. If the dolphin touched a symbol, it would be given the corresponding object — a ball, a ring, or a float necklace. The question: Would the dolphin learn to use the keyboard to get the preferred toy? I hypothesized the answer would be yes.

This style of interspecies communication had been developed with chimpanzees, but no one had attempted it with dolphins. As simple as this transaction sounds, it does in fact require considerable cognitive abilities.

The prospects for doing all this in Paris were not encouraging.

With Busnel's help I looked farther afield. A small aquarium, the Zoo Marin at Port Barcares, soon popped up on our screen, and I visited the place in November 1978 and met the young female dolphin that had recently arrived, captured from the wild. (Given what we know today, I am strongly opposed to the capture of dolphins from the wild for any reason.)

There were three bottlenose dolphins in the too-small pool: a large male, called Hoss, a smaller female, Niki, and a young female that I guessed was about three years old. She was smaller than her companions, and although she had arrived six months before, she appeared reticent around them, especially Hoss. I watched as the three swam unhurriedly around the tank, the newcomer always staying quite close to Niki. "If she looked at me," I wrote in my notebook, referring to the newcomer, "I gently tapped [my open hand on the side of the tank] three times. It brought her closer. As she was close, I tapped three times again. Finally, in 20 minutes she had [tentatively] touched my hand. I felt it was a good indicator — and I decided to come and work with her."

Monsieur Stone readily agreed that I could spend six months there doing my graduate research, but there was one condition: I had to teach the young dolphin, who I now called Circe in my mind, to come to a particular spot by the side of the pool and be fed, part of preparing her to be in a dolphin show with her companions. Dolphins are predators, and their natural diet is live fish. At most aquariums, and in this case, frozen and then thawed fish are substituted for the real thing. Nevertheless, dolphins soon adapt, and the fish is nutritious. Despite my distaste for using dolphins for human entertainment, I accepted the arrangement as a compromise that allowed me to begin my work.

I rented a small apartment in a summer resort complex composed of several boxlike, stucco buildings on the beach with a spectacular view of the sea, directly across the road from the Marine Zoo. This being January, I was the sole resident apart from the manager and his family. Each building was painted a bright white, so they all stood in jarring contrast to the natural browns and grays of the Pyrenees and the azure of the Mediterranean. The day after I arrived in Port Barcares I went to see Circe and the other dolphins, and late that afternoon, as I walked back to my apartment across the seashell-strewn plain that separates the foothills of the Pyrenees from the sea, I felt myself drawn to an area to the right of my path. I didn't know why. Then I noticed a tuft of fur quivering slightly in the grass. I walked toward it, knelt, and saw a tiny baby rabbit, with a skein of fur all but ripped off its back, hanging loosely; its skin was shiny, but there was no blood in sight. I scooped up the poor thing and nestled it against my neck inside the thick green French army parka that I was wearing to ward off the January cold.

I got back to my apartment, warmed some milk, and fed the bunny as best I could. I checked it for bugs (I am practical as well as compassionate!), and then I retired to my narrow bed, the little creature tucked against my neck, for extra warmth.

I woke the next morning to the sight of the bunny hopping around on my bed, its fur positioned where it should be on its back and looking fine. It had survived! So began a new, but temporary, relationship. I built the baby rabbit a hutch inside the apartment, and when the weather warmed sufficiently, I created an outside space for it too. It became my companion for the next few months. I eventually found a permanent home for the little guy, with a local family who loved animals and thought of rabbits as pets and not as a next meal. Somehow it seemed a good sign for the work to come. We scientists can be a superstitious bunch.

Meanwhile, my formal work had not started well. "The electricity has been out since the day after my arrival," I wrote in my log on January 24, 1979. "So the experiments must wait until the emergency situation is over." So I began building a relationship with Circe, who was quite timid with me at first, and started to work on the feeding regimen. Circe had seemed a little intimidated by Hoss and to a lesser extent by Niki when I'd met her the previous November, and she still was. I spent time just observing Circe's behavior and that of the other dolphins to see if they would respond as Dal and Suwa had. "Niki was near Skip [a sea lion in an adjacent enclosure] the whole time," I noted in my log one night after work. "Circe and Hoss were by me the whole time — floating — rolling etc. At one point I stopped, Hoss began vocalizing on a note. I answered. He answered — changing as if going through his different combination of sounds. Then he stood on his head — keeping his tail out of the water . . . Circe just watches." She seemed nervous to me, and she barely ate anything at suppertime. "She fears H I think," I wrote in my log. "We need our own space."

I continued to try to build Circe's fragile confidence over the next few days but still wasn't able to do any serious work. "I've found a quiet moment to sit and write, finally," I noted in my log three days later. "The electricity is still stopped. I'm beginning to worry if it will ever come on again." I was encouraged that Circe seemed to be getting the hang of coming to station to be fed, but she was very picky about what part of the fish she would eat. She seemed to prefer the heads.

Five days into the project, Circe and I had our first intimate moment, tentative on her part. I had noticed that after she ate, she seemed to like to have some time to herself, five to ten minutes. Then she would come over to me, without my having to call her. On this morning, she came very close to me, looked at me, and seemed to want me to touch her. I put my hand in the water. "Finally she touched," I wrote in my log that evening, "first with her mouth closed, putting the very tip of her head against my hand. Then she began gently pushing my hand up to the surface. At one point she kept opening her mouth under water." When working with dolphins or any other animals, pets included, you have to "read" their behavior, their body language. Circe's body was relaxed and her mouth was open in a calm manner as well — I read her signals as friendly and solicitous, and I acted on this interpretation. "I put my hand on her mouth to show her I trusted her. I could run my fingers along her closed mouth gently, and over her rostrum — & back. Finally

I put my hand in her open mouth and it was gentle, relaxed and open. I tickled her tongue, she moved closer and closed her eyes." This little incident marked a turning point in our relationship.

The lack of electricity at the facility was increasingly frustrating, not least because I had to haul buckets of warm water from my apartment, the only place that did have electricity, to thaw the fish for the dolphins. On February 1, more than a week after the outage began, the electricity was restored, "thanks to prayers — sympathetic magic — chanting — electricity dances, etc." At last, I could begin.

I started informally. I wanted to give Circe more choice and control over getting objects that she desired. I first taught her that she could obtain a toy by simply touching it when I presented it to her at the side of the pool. She learned this quickly, and apparently getting the toys themselves was reinforcing.

Then I conducted my first formal experiment: testing if a dolphin could visually discriminate between the white symbols. This was not a trivial question at the time; the everyday world of dolphins was thought to be dominated by sound, not sight, and, as I mentioned before, dolphins' echolocation abilities are exquisite. I therefore couldn't *assume* that dolphins could see well enough and discriminate well enough to use the visual symbols of the keyboard. Back in 1979 when I started with Circe, we did not know if they could. More than a decade earlier, at the Stanford Research Institute in Menlo Park, Winthrop Kellogg and Charles Rice had done some very preliminary studies on this kind of ability in dolphins, but they had gone only so far. Essentially, I was starting at square one. The first part of my doctoral research, therefore, was to discover whether dolphins could visually discriminate among simple shapes.

I built a simple wood-and-metal apparatus that displayed a set of white, wooden visual forms: a circle, a cross, and a triangle. The forms were buoyant and backed with metal studs so that they could float and adhere to the apparatus. The apparatus consisted of a horizontal bar with a vertical bar coming down off each end. This was the setup for a match-to-sample task: one symbol (the sample) was displayed in the center of the horizontal bar, an identical (matching) symbol was on one vertical arm, and a different (nonmatching) symbol was on the other. A symbol was introduced as the sample, and Circe had to look at the sample and pick the correct match. Dolphins are adept at touching things with the tip of their rostrums (the front of their mouths), and Circe was no exception. The symbols were presented in air so she had to use vision, not echolocation, for the task.

Using this setup, I rewarded Circe with a piece of fish when she touched the bottom circle, the correct match to the sample, but not when she chose the triangle. I switched the positions of the circle and triangle at the bottom from time to time, in what we call a pseudo-random pattern, to make sure she wasn't using the symbol's position or any predictable pattern to solve the problem. Other animals are clever like us and look for useful strategies to solve problems, like choosing what's on the right because that's what worked last time or trying an alternating pattern such as right, then left, then right. Circe was an eager student, and she learned the procedure fairly quickly. So now I knew that, indeed, Circe was able to identify shapes visually and could compare and distinguish between different forms.

The next procedure built on Circe's previous understanding of the match-to-sample task. The conditional-discrimination task was a bit more of a cognitive challenge, one that David Premack had pioneered with his chimpanzee Sarah. The task involved learning to associate a non-iconic visual symbol — that is, a symbol not visually similar to what it will be linked with — with a particular object; for example, a triangle with a toy ball. In this case, using the same match-to-sample apparatus as the first experiment, I put a ball in the top position on the keyboard, the sample position. In the two lower positions I had a triangle and a cross. For this test I rewarded Circe when she touched the triangle key, but not when she touched the cross; I did the same type of thing with the other objects. Again, Circe got it quickly, showing that, indeed, she had learned the concept of conditional discrimination. In the most parsimonious explanation and the driest technical description, what Circe learned was this: If the ball was in the sample position, then she should touch the triangle; if the ring was in the sample position, then she should touch the cross; and if the necklace float was in the sample position, then she should touch the circle. A more user-friendly description of Circe's newly acquired ability was that the triangle was associated with a ball. It is too much of a stretch to infer that in Circe's mind, the triangle was a "word" for ball, at least not in the way we understand and use words. But this was what I hoped my work would ultimately lead to, the development of an artificial language that humans and dolphins could use to communicate.

I took Circe a step farther down this cognitive path with the third experiment, preference testing and free choice. First, I put the three toys — a ball, the necklace, and a ring — in the pool with her. I then sat by the pool for several days, notepad in hand, carefully recording how often she played with each of the toys. I wanted to know which one she preferred most. Ball was her favorite, followed by necklace, followed by ring. Most dolphins love to play with toys, but Circe was especially playful. She would push or carry the toys around the pool or toss them into the air, and often she tossed one of them to me. I tossed it right back, so this piece of work was essentially a game, one that strengthened the bond that was developing between us.

The real challenge came in the next part of the study. Circe had learned a conditional association of a specific object with a specific symbol: the triangle with the ball, the circle with the necklace float, and the cross with the ring. So now we were all set to do the test and see if Circe would use the symbols to obtain toys and if the symbols she used would match her toy preferences. I put the three symbols on the keyboard, positioned them so she would have equal, easy access to all three positions, and then watched what she did. Would she touch the triangle more than the circle, and the circle more than the cross? This was new cognitive ground for dolphins, and Circe was a star. Given her free choice and control over the keyboard, she asked for her most preferred object most often, the next preferred object a bit less often, and the least preferred object least often. In scientific terms it was a beautiful linear correlation suggesting that Circe could ask for an object she wanted.

Recently I was looking at one of my old notebooks from the lab in Paris, and I came across the proposal I had written for my future research plans. It was: "To create a third language or communication system to exchange information between dolphins, *Tursiops truncatus,* and humans, *Homo sapiens.*" I couldn't help laughing as I read these lines. It is all very lofty and PhD-student-like, very earnest. You have to remember that I was deeply immersed in communication theory and cybernetics, a systems approach to communication. Very mechanistic in a way, looking at the different parts that interact in the system: a human, a dolphin, and a keyboard, which is the interface. "The human experimenter" — that would be me! — "will begin by establishing a relationship with the dolphins," the proposal continued, "establishing trust, and mapping out environment . . ." And so on.

At the time, John Lilly expected he'd be able to teach dolphins to communicate with humans in English, which I considered to be unrealistic and less interesting than understanding their own forms of communication. While not as extreme as Lilly's notions, my proposal was definitely out there, even naive in its wide-eyed expectations. Nevertheless, in those months working with Circe during the late winter and through the summer of 1979, in the shadow of the majestic Pyrenees, I took the first steps down that ambitious road. I had my first real glimpse into the dolphin mind.

But there was more than that. I had really bonded with Circe, and I think she had with me. I was always eager to see her, and she always greeted me with excitement when I approached the pool, rapidly swimming around, occasionally porpoising a few times, then coming to where I stood, eyes big and eager looking. I rubbed her head, and her belly when she offered it. I didn't go into the water with her, partly because it was very cold in the first months, and then because Hoss's presence made it potentially hazardous. But despite the distance between us, me on land and Circe in the water, a bond had been created and I was entangled with her. I worried about her future and I anticipated that my departure from her would be difficult.

A few months later, I had to leave Port Barcares for good, to return to Paris and then the United States, and for the first time I experienced the emotional rawness of the loss of a very special relationship; a grieving, really. My time with Dal and Suwa in Florida two years earlier had been short, and in any case, their emotional allegiance had been to Betty, not me. Leaving them was relatively easy. So I wasn't prepared for the pain of saying goodbye to Circe, who had certainly enchanted me. I wasn't prepared for the agony of not being able to explain to her why I was leaving. I wasn't prepared for the searing feeling of guilt about abandoning her. I left, expecting I'd see her again one day. I never did.*

⟡

Sometimes in science the most jaw-dropping insights come when you are not even asking a profound question. So it was with Circe and what I call the time-out story. The circumstances could hardly have been more mundane.

You will remember that part of my deal in working at Monsieur Stone's marine zoo was that I would teach Circe to come to a particular spot at the side of the pool and stay at a particular location when being fed. Even smart animals need to be trained for such husbandry practices. But just a few weeks into this little train-

---

* I tried to locate her later that year, but she had been moved to another aquarium in Europe, and no further information was available.

ing program, Circe gave me a glimpse into her mind that revealed just how *very* smart she was.

I was feeding Circe Spanish mackerel; since they were too big for her to swallow whole, I cut the fish into three smaller sections: heads, middles, and tails. After only a few days, Circe had learned to come to the side of the pool and station (stay in position in front of me) while I fed her. I used basic operant-training techniques to teach her to eat the various sections of fish and to stay at station; that is, I rewarded her with fish and vocal praise for doing the right thing, approaching the side of the pool when given a specific hand signal and then staying in front of me. She had to learn not to leave station until I gave her a signal that meant the session was over. If she left prematurely, I had to communicate to her that it was the wrong thing to do, so I'd give her a time-out, just as parents use time-outs with their children to let them know they've done something wrong and to allow them time to think about it. I'd simply break social contact and move away, leaving her alone for a short while. Time-outs can be extremely brief; just stepping back or turning away for a moment can let an animal know that it has done something wrong, and the animal then has to wait a short time before it can obtain another reward. (Nowadays, time-outs are used less frequently with dolphins because they can lead to frustration on the part of the animals.)

So, I used a time-out with Circe in the first days of teaching her to station in feeding sessions. I would back away from the side of the pool and just stand there looking at her for a few seconds to a few minutes. Then I would return and continue the session. Circe would always come right back to the side of the pool and wait for me to return. The time-out procedure worked well, and Circe soon learned to stay at station. Circe was a good eater, and she usually ate all the food offered; if she didn't want to eat all the fish, that was okay, and I interacted with her at station. But she had to stay with me until the end of the feeding period while the other dolphins were being fed.

During feeding sessions in the first week, Circe readily ate the fish heads and sometimes the middles, but I noticed that she was spitting out the tails each time. I thought that maybe she didn't like the tails because of the spiny fins. So, trying to please her, I began cutting off the tail fins. It worked. She immediately ate all the trimmed tails! I remember laughing and thinking to myself, *Circe is training me to cut her fish the way she likes it!*

It was two weeks after I first started using the time-out procedure, and everything was going quite well. I was learning about Circe's patterns of behavior. I noticed she was also frequently watching me as well. One day during a feeding I accidentally gave her an untrimmed tail. She immediately looked up at me, waved her head from side to side with wide-open eyes, and spat out the fish. Then she quickly left station, swam to the other side of the pool, and positioned herself vertically in the water. She stayed there against the opposite wall and just looked at me from across the pool. This vertical position was an unusual posture for her to maintain.

I could hardly believe it. I felt that Circe was giving me a time-out! She stayed there for a short time, vertical, looking at me, and then returned to where I had been waiting at poolside. I continued the feeding, and she ate all the properly cut fish and never broke station again during the session.

Did Circe really give me a time-out? Did she really intend to correct my behavior or let me know that I'd done something wrong, just as I had done with her? It felt like that to me. But, as a scientist, I knew that I needed more information to be sure. Impressive though it was, this one case was not enough. It was just an anecdote, and maybe I was reading too much into it.

Circe's unusual behavior prompted me to do an experiment. I waited for a few days, during which time I continued to cut her fish as she liked it, and she ate all the fish and never left station. Then one day I deliberately offered her another untrimmed tail. But this time I planned it as an official experiment to test my idea. I had an idea, or hypothesis, and I set out to test it. This is the basic scientific method. And guess what: Circe gave me another time-out. Over the next weeks I repeated this procedure, six times in all, and got the time-out response on four of them.

This is a powerful example of how communication develops between people, and between people and other animals. We all have cultural or species-specific signals that are used to transmit, receive, and share information with others. In social interactions, we observe and learn about others' patterns of behavior and signals. Communication is a social phenomenon, and all animals communicate. There are two basic tenets of communication theory: First, meaningful communication develops through social interactions. Communication is a dynamic transactional process of sending, receiving, exchanging, and interpreting signals; we interpret signals and messages based on our expectations, which are shaped by our past experiences. Second, there is no absolute meaning in messages — meaning resides in the individual's interpretation of messages. There's no way to know exactly what a word or signal means to another individual at any given time, but if the communication signals work pragmatically (practically) to serve a particular function, participants will use signals to obtain certain results. So we communicate — it is a matter of developing a synchrony of behavior between two or more participants. Circe's use of the time-out was a superb example of how the behavior of two organisms (a human and a dolphin) becomes synchronized and meaningful to each through social interactions.

It seemed to me that Circe understood that when I gave her a time-out, it meant that she had done something wrong. So when I did something that she considered wrong, giving her unwanted untrimmed fish, she used the same signal, a time-out, to communicate that to me. This was my first real insight into the depth of the intelligence of this remarkable and highly social species. You will hear a good deal about what is called theory of mind in the following pages. It means, very simply, that an individual is able to imagine the mind of another and understand how that other will behave. In the vernacular, one might say that the first individual is reading the mind of the second. Theory of mind is a highly sophisticated cognitive ability, one that is very rare in the animal world (humans and great apes have the ability). This time-out event with Circe demonstrated that she and — by extension — other dolphins are members of that elite club.

I eventually included the time-out experiment as part of my doctoral thesis on dolphin communication.[3] It was, to me, a pattern that communicated, a pattern that connects.

3

IN SEARCH OF THE DOLPHIN
ROSETTA STONE

IT ISN'T ALWAYS easy to know when a pregnant bottlenose dolphin is about to give birth. But I was going to find out, or so I hoped. It was late July 1983, a little less than a year since I'd set up a new research program at Marine World Africa U.S.A. in Redwood City, about forty minutes south of San Francisco. The three dolphins I observed were a somewhat unusual social group: two female Atlantic bottlenose dolphins, eighteen-year-old Terry and eight-year-old Spray (I soon renamed her Circe in honor of my first dolphin mentor, whom I still held dear), and a huge, older male Pacific bottlenose dolphin, Gordo.

Although I didn't know it when I arrived at Marine World that fall, both the female dolphins in the research pool were pregnant. This was an unexpected bonus, because it offered me the opportunity to observe the development of vocalization, echolocation, and social behavior of young dolphins right from birth.

My immediate issue on that last day of July was whether Terry, the older of the two mothers-to-be, was going to calve any time soon. About a month earlier, Terry's belly region had become quite distended, a sign that birth was not far away, a matter of weeks. My coworkers and I had seen her flexing and crunching occasionally over the previous few days, behavioral signs of an impending birth. That told us that delivery was imminent. Everyone at the pool was in a state of high alert, and more than a little anxious. During the previous few years there had been several births among the performing dolphins, but for various unrelated reasons, none of the calves had survived. I was determined that Terry's and Circe's infants weren't going to suffer the same fate, especially as these were my first dolphin births. Drawing on my experience as a set designer, I constructed a playpenlike contraption around the research pool that could be hoisted into place at night. I wasn't going to have any baby dolphins flying out of the pool on my watch!

It was early evening on July 31 and I had been at the pool for forty-eight hours straight. When I could, I slept on a cot in the research lab, a small eight-by-twelve-foot office perched on a grassy mound about fifteen feet away from the dolphin pools. My staff and I anxiously watched and waited for Terry to go into full labor. I was exhausted, and I needed to get a change of clothes. Terry had been off her feed earlier in the day, which is also sometimes — but not always — a sign of imminent delivery. I discussed the options with my research assistant Bruce Silverman. Did I have time to run home, check on my two cats, who were being fed by my neighbor, and grab some fresh clothes? Or would it be better to wait and go tomorrow? Of course neither of us knew for sure. Terry was still flexing and crunching, but not as often as she had over the past few days. We figured it would be at least another day, maybe more. "Okay," I said to Bruce, "I'll risk it. I'm going to the city. Call me if anything develops."

I jumped in my car, and in forty-five minutes I was at the door of my apartment in San Francisco. I could hear the phone ringing. I fumbled with my keys, flung the door open, and snatched up the phone. It was Bruce. "Terry is giving birth right now," he said. "Come quickly or you'll miss it." I grabbed a change of clothes and raced back. When I got there, Terry was swimming calmly around the pool, solicitous of her newborn calf at her side. I had missed the birth!

Terry had given birth before, and she looked like a competent mom and knew exactly what to do. A good thing, because newborn dolphins are a little unsteady, as all newborns are, even ones as precocious as these. They come into the world after a twelve-month gestation period, and they must immediately go to the surface to breathe and then be able to swim well enough to keep up with their mothers. They are the aquatic equivalent of newborn foals, a little wobbly on their (metaphorical) feet. Terry guided her newborn calf by steering his swimming direction with her whole body, occasionally stroking him with her pectoral fin, an act of maternal comfort and reassurance. At one point the calf veered off course and rammed into the side of the pool. Terry was immediately at the calf's side. I heard calls that I later understood were distress whistles. Was it Terry or her calf? I didn't know.

Terry seemed to be actively adjusting the calf's breathing rate by occasionally putting her rostrum over his blowhole. Very soon the calf's breathing rate was down from an initial eight or nine breaths a minute to a more natural three to five. Within a few hours the calf was interested in nursing, following his instincts to search for the creases that concealed his mother's nipples, another anatomical adaptation for a streamlined body. The little guy blundered around awkwardly at first, bumping against Terry's body, trying to insert his thin tongue into the crease at the corner of his mother's jaw and then into the crease where the back edge of her pectoral fin met her torso. Terry patiently rolled over on her side and oriented her body so that the calf's rostrum was next to her mammary slits, which are on either side of the genital slit, near the tail. He immediately slipped his tongue into the crevice, formed a watertight seal, and began to suckle. We breathed a sigh a relief. The calf had passed his first hurdle. Before long the two folds across his body, the fetal folds, would disappear as he put on blubber fueled by his mother's rich milk.

Meanwhile, Circe was showing signs of imminent delivery. She had begun flexing and crunching just before Terry had given birth. Circe was much younger than Terry and had not given birth before.

So this was to be a first for both of us, because I was determined not to miss this one. I was going to stay on-site until the calf was born, however long it took.

I didn't have to wait that long. The morning of the second day after Terry gave birth, Circe was off her feed. We were ready. We had video cameras and audio recorders running continually, and my research staff was monitoring Terry and her calf, making sure that things were going as they should. I sounded like a play-by-play sports announcer as I recorded minute-by-minute commentary of Circe's every move. By early evening Circe was swimming in tight circles, sometimes doing corkscrew movements in the water, and looking a little frantic. Terry, who'd been close to her calf constantly since his birth, left him for a few seconds to swim alongside the apparently anxious Circe. Then Terry resumed her vigilant mothering of her own calf but continued to watch her companion from a distance. Finally, I saw the first real sign of birth, the appearance of the baby's tail fluke from the birth canal.

Some days earlier, we had made an important and potentially disastrous decision. Aquarium practice is to keep male dolphins away when females give birth because the males can sometimes be aggressive to newborn calves. The male in this case was Gordo, father of both Terry's calf and Circe's yet-to-be-born calf. He was the gentlest of dolphins, a real sweetheart. It would be difficult to incorporate him into another social group at the aquarium; he was an older male dolphin, and the change would have been very stressful. So I followed my gut and let him stay. Now, as Circe's first birthing began, Terry herded Gordo to the side of the pool, literally pushing him against the pool wall. She positioned her infant between Gordo and her own body, right below where I was standing near the pool. Terry and I both watched Circe. I could see the baby's tail fluke moving back and forth a little, peeking out of the genital slit. Then more of the tail, the tail stock, emerged. All of a sudden, with a burst of blood, the baby was out.

The baby bobbed about in the water, obviously in need of help.

Circe, who began swimming aimlessly around the pool, looked in need of some help as well. There was quite a crowd of us watching, including our veterinarian and many of the research and marine mammal staff, and we were all thinking, *Circe, go and get the calf!* Dolphin lore holds that females are hard-wired to push their newborns to the surface, a supposedly powerful instinct that some believe underlies the instances when dolphins save drowning sailors or others in trouble in the sea. Yet it appeared to us that Circe (like many other dolphins I've since observed) had no instinct to do any such thing and that she was completely bewildered. I was listening to the dolphins on my headset via hydrophones; the pool was strangely silent. Terry, still pressing her calf against Gordo and pinning them both to the side of the pool, had her head turned at almost a right angle as she looked at Circe. Suddenly, Terry produced a long, complex whistle that began and ended with her personal contact call, a whistle a bit different in unique ways from the basic contact call of each of the others in the group, used as if to signal *Hey, it's me!* and perhaps even more. Circe instantly snapped into action: she immediately swam to the floundering baby and started to tend to him just like any good dolphin mother. Terry's whistle had been a combination call, a stringing-together of her contact call and a series of sounds that meant nothing to us but clearly meant something to Circe.

Had Terry recognized Circe's inexperience and bewilderment and given Circe instructions on what she was supposed to do? I still can't say, but it certainly looked that way. In any case, the two mothers and their babies began to swim together, beginning a companionship that was to last for a very long time. Continuing my tradition of naming dolphins after the Greek gods and goddesses, I named Terry's calf Pan and Circe's calf Delphi. Pan and Delphi became "the boys," as we fondly called them.

◦◦◦

At the time Pan and Delphi were born, I was founder and director of Project CIRCE — the Cetacean Intelligence Research and Communication Experiment. I had completed my doctoral research with the first Circe in France in June 1982 and then returned to the States, where I spent the summer months writing up my dissertation. After receiving my PhD, I wanted to build the underwater-keyboard system that would, I hoped, enrich the lives of captive dolphins, providing them with increased choices and control over their toys and activities while at the same time enhancing our scientific knowledge of their cognitive abilities. I had already envisioned the keyboard: it would hang vertically before the dolphins with visual symbols on its face. When the dolphin pressed its rostrum to a specific key, it would hear a specific computer-generated whistle and receive a specific object or activity. Quite simple, really — like a dolphin vending machine, but one that would work underwater and would interface with a computer in the lab so I could collect and track the dolphins' use of the keyboard over time. I was eager to get started and I wasn't interested in doing the usual postdoc stint in another lab. I was ready to start my own work, but that required the right social group of dolphins. I would have much preferred to study dolphins in the wild, but this work had to be done with dolphins in an aquarium setting. I had to get up close and personal with the dolphins and run controlled studies that could be carefully documented and analyzed. Someday I hoped to do fieldwork to learn about the secret social lives of wild dolphins. Yet even just spotting dolphins in the wild was difficult, like looking for a needle in a haystack, as field researchers knew. Like astronomers searching the skies for undiscovered stellar bodies, one searches the glimmering waters in the hopes of spotting triangular dorsal fins at the surface. It is both thrilling and frustrating.

In what I can only recall as an unbelievably odd set of circumstances, I visited Marine World and somehow convinced the owner of the facility, Michael Demetrios, to convert what had been a petting pool into a research pool. In retrospect this was an amazing feat, because petting pools were a big attraction. (Hopefully, given our increased understanding of and respect for dolphins, petting pools soon will be — and *should* be — a thing of the past.)

My keyboard research — focused on Pan and Delphi — had quite ambitious goals. I had three main questions. The first and most basic: What would happen if we gave dolphins choice and control by way of a keyboard? The standard research approach at that time used techniques that were similar to those used when training dolphins to perform specific behaviors in shows. Researchers gradually shaped, or modified, dolphin behavior through reinforcement. Effectively, a trainer or experimenter decided what response or behavior he wanted from an animal and then selectively rewarded that response with food. The goal was to limit the dolphin's freedom so that when the animal was given a specific cue, such as a hand or acoustic signal, it would respond in one way. In these cases, it is the experimenter or trainer who is in control.

In contrast, my goal was to abdicate power and turn at least some choice and control over to the dolphins and see what their big brains would do with it. My assumption was that if you treated another as an intelligent being, he or she would show you a reflection of that intelligence.

My second question: How do dolphins learn their whistle repertoires? I had embarked on a project to decode whistle repertoires by observing and recording the behavior of the group. Vocal learning is surprisingly rare in the animal world, and besides humans, only avian species, dolphins, whales, and possibly elephants and bats show evidence of it. Numerous anecdotal accounts reported dolphin vocal imitation, yet no studies existed that investigated the process or extent of their learning, apart from Lilly's attempt to teach them to speak English using food rewards. I wondered how dolphins learned from one another without any human trainers. Was it at all similar to how humans and birds learned, through exposure and through imitation and practice?

Two decades after Lilly's philosophy had been most popular,

Dolphin mythology: Neptune, crowned with laurel, offers a dolphin. NYPL

Dolphin coin from ancient Thrace, later 5th–4th century B.C.E. Poseidon on one side of the coin with Eros riding a dolphin on the flip side, c. 300 B.C.E. AUTHOR

Coins in the form of leaping dolphins were common throughout ancient Greece and many other countries. This coin was used on the shores of the Black Sea. In ancient times and today, the Black Sea is home to bottlenose dolphins. AUTHOR

Circe, my first dolphin mentor in France. AUTHOR

The newborn calves, Delphi and Pan, accompanied by their mothers, Circe and Terry. AUTHOR

Delphi and Pan,
a close alliance.
AUTHOR

Hand to pec rubbing with
Delphi and Pan. AUTHOR

Staring eye-to-eye with Delphi, encountering a strong presence. AUTHOR

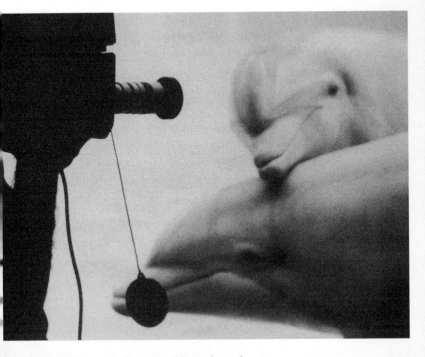

Dolphins looking in camera, observing and being observed. AUTHOR

The underwater keyboard provided the dolphins with choice and control. When Delphi and Pan pressed a key they heard a computer-generated whistle and got specific objects or activities. AUTHOR

Reiss at keyboard rubbing Delphi after he presses "rub" key. AUTHOR

Reiss gives Pan a ring after he hits the "ring" key. AUTHOR

Presley with a nontoxic
mark on his head in the
mark test. AUTHOR

Presley at the mirror viewing himself and
passing the mark test. AUTHOR

Reflecting on himself. AUTHOR

Dolphin admiring her own creation, an exquisite bubble ring. AUTHOR

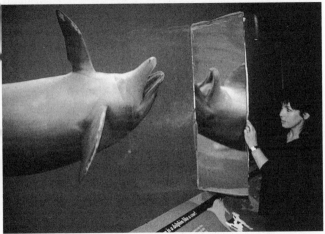

After the MSR paper was published, the dolphins continued to get mirrors as enrichment objects. AUTHOR

One of the first hurdles in the rescue. Tense times as Humphrey stopped and refused to swim through the pilings at the Liberty Bridge.

ASSOCIATED PRESS

Humphrey follows the *Bootlegger* back out to sea. ASSOCIATED PRESS

Despite press conferences (here, at the National Press Club, with leading marine biologists and veterinarians), diplomacy, and public pressure, the drive hunts continue. AUTHOR

After the film *The Cove* was released, the dolphin slaughtering was moved under tarpaulins, to try to hide it from any rogue cameras. HELENE O'BARRY/DOLPHIN PROJECT

expectations of the cognitive reach of dolphin minds were both lesser and greater than they had been.

During the first half of the twentieth century, the scientific study of animal psychology was dominated by a philosophical and theoretical approach known as behaviorism. Put simply, the main tenet of behaviorism is that scientific investigation should focus only on animals' (including humans') objective, observable behavior and their reactions to environmental stimuli. Behavior in animals was the outcome of stimulus/response processes, and nothing more. Mental processes might underlie some behavior, but because they could not be observed directly, they should not be the subject of scientific investigation.

In other words, animals were considered little more than biological automatons, creatures that lacked emotions or intent. Mental processes such as reflection and self-awareness, or consciousness, were nowhere in the picture, the behaviorists argued. This approach influenced the methods by which psychologists and biologists studied animal behavior. The brain was a black box, unknowable to human observers, and therefore not a valid topic for scientific inquiry.

It was not always like this. Charles Darwin would have been shocked at such a mechanistic perspective of the animal mind. In 1872, a decade before he died, Darwin published a book called *The Expression of the Emotions in Man and Animals,* which presented the then commonly held belief that humans were not the only animals to experience thoughts, plans, and emotions. But behaviorism pushed this idea aside until well after World War II. Beginning in the 1950s, however, there was a growing interest in the nascent fields of artificial intelligence, computer science, and neuroscience. Two landmark scientific conferences in 1956, one at the Massachusetts Institute of Technology and one at Dartmouth College, marked a paradigm shift in thinking about minds, brains, and computers. These meetings resulted in what has been termed the cognitive revolution in how scientists approached the question of thinking and the minds of humans and other animals. The fundamental question changed from *Can* other animals think? to *How* do other animals think?

The modern study of animal minds was largely influenced by a small but courageous and pioneering book written by Donald Griffin in the late 1970s. Griffin had established his reputation three decades earlier; in 1944, at the age of twenty-nine, he and his fellow student Robert Galambos discovered that bats navigate using sonar, which they called echolocation. This was a revolutionary suggestion back then, and it was met in some quarters by complete disbelief. Echolocation was eventually accepted as a biological reality, of course, and it was followed several years later by the discovery that dolphins also employ echolocation for navigation and detection of prey, although in water rather than air.

In 1976, Griffin published *The Question of Animal Awareness,* which argued that nonhuman animals might also experience consciousness and be capable of thoughts and intent. He faced disbelief once again, and sometimes contemptuous opposition. Nevertheless, the seeds of what would become known as the field of cognitive ethology had been sown. Exploring animal minds became a more respectable scientific endeavor. The field is still criticized for citing anecdotes — which are subject to interpretation — as evidence. The late Stephen Jay Gould often said that "the plural of anecdotes is data." Yet as Marc Bekoff, a professor emeritus of ecology and evolutionary biology, noted, anecdotes, like anthropomorphism, "can be used to make for better science, if we only let them."[1]

By nature, I have a lot of sympathy with Gould's and Bekoff's viewpoints, as long as the anecdotes are combined with scientific rigor. I try to trust my instincts, trust the insights that emerge, but I verify them with scientific tests as meticulous as the circumstances allow. By the time Pan and Delphi were born, dolphins were recognized as intelligent, sentient creatures. But to what degree? That remained to be discovered. John Lilly's claim that dolphins could produce human words had proven to be unfounded. Yet they communicated with one another very effectively.

❧

Animal research methods were still hotly debated at this time. In 1980, I attended a conference at the prestigious New York Academy of Sciences entitled "The Clever Hans Phenomenon: Communication with Horses, Whales, Apes, and People," ostensibly about current research on human-animal communication, specifically apes and dolphins. As a neophyte in the field, I was very eager to attend. I was aware that so-called ape-language research was under growing scrutiny, and I wanted to hear more about that. From the very first moment that the conference opened, I found myself witness to a vicious and vitriolic attack on the integrity of ape-language researchers in particular and the field of interspecies communication research as a whole. It made me wonder not whether humans could communicate with other animals but whether humans could communicate with other humans. I spent much of the time with Irene Pepperberg, who was at the beginning of what turned out to be a remarkable research program with an African Grey parrot named Alex. Irene and I felt like a couple of kids who had inadvertently stumbled into a roomful of adults slinging insults at one another.

Under attack were ape-language researchers including Allen and Beatrix Gardner, David Premack, Roger Fouts, Duane Rumbaugh and Sue Savage Rumbaugh, and Penny Patterson. They had used sign language, keyboards, and other kinds of symbols to explore whether apes could produce "words" in relation to objects or behaviors and whether they had a grasp of syntax. Only Sue Savage Rumbaugh and Duane Rumbaugh attended the conference.

Those doing the attacking included experts in training circus animals and in the arts of trickery of various kinds. One critic was the Amazing Randi, a professional magician famous for unmasking self-proclaimed psychics. It was, wrote a reporter for *Science*, "a celebration of deception in all its varieties."[2] The title of the conference, "The Clever Hans Phenomenon," referred to a famous German horse that performed arithmetical tricks at the turn of the twentieth century. His owner, Wilhelm Von Osten, a retired high-school math teacher, claimed he had trained Hans to add, subtract, divide, multiply, and perform other mathematical and intellectual feats. He would ask Hans questions that required numerical answers; for instance, he'd ask, "If Friday is the fourth of October, what is the date on the following Tuesday?" Hans would then tap his hoof eight times, to the great amazement and delight of the audience. The secret behind this apparent phenomenon was this: as Hans approached the correct number of taps, his trainer, unwittingly, tensed up; when Hans reached the right answer, his trainer very slightly relaxed, displaying relief in his imperceptible posture change. Imperceptible to ordinary folk, that is, but not to Hans, who was able to detect head movements as minute as a fifth of a millimeter. The trainer was, unknowingly, cuing Clever Hans — Von Osten wasn't deceiving his audiences; he was deceiving himself.

Clever Hans's talent was not in mathematics but rather in his extraordinarily keen perception. When Hans was tested in the absence of Von Osten, he could still solve the problems, provided the person testing him knew the correct answer; Hans read the subtle, inadvertent cues of his questioner. This is known as the Clever Hans phenomenon. Scientists now recognize the need to eliminate their own unconscious cues when they're testing animals, but they also recognize the remarkable sensory acuity and problem-solving abilities of their subjects. Clever Hans was clever indeed — but not in the way people thought.

Essentially, the meeting addressed the possibility that the ape-language researchers, and others doing similar work on different species, were, like Von Osten, deceiving themselves. In truth, the people involved in ape-language research at that time had done some very powerful studies and had accumulated persuasive evidence that apes (and an African Grey parrot) could indeed learn to associate symbols with objects and actions. Whether one would call that language is another matter. In any case, they were well aware of the dangers of cuing and were more sophisticated than their critics in their understanding of what communication was all about.

Dolphins can be terrifically subtle readers of cues. At the conference, John Prescott described a dolphin at the New England Aquarium who had been trained to respond to a visual signal, an arm raised in a particular direction, by leaping out of the water and jumping over a bar. In an experiment, the dolphin was able to do this same behavior *even when it was blindfolded*. The trainer assumed that the dolphin accomplished this by using echolocation through the air as a substitute for visual cues. Dolphin echolocation is usually done underwater, of course, yet this seemed to be the only explanation.

It was, however, wrong. Prescott's colleagues eventually realized that when the trainer raised his arm as the signal to jump, he preceded it by taking a half a step forward. Unbeknownst to the trainer, that step provided an auditory cue to the dolphin, which, like all dolphins, had very acute hearing. It was the sound of the step, not the arm raising, that the dolphin understood as the signal to jump. When the trainer performed the signal he had been using for seven years with this dolphin but without taking that little step, the dolphin simply remained in front of the trainer and did nothing.

Karen Pryor was at the conference too. I often cite her best example of cuing, which occurred at Sea Life Park Hawaii. One day one of the trainers came to her, visibly excited, and said, "The dolphins are psychic! This particular dolphin is psychic!" Karen asked her what on earth she meant, and the young woman explained that the dolphin would obey a gestural command before she actually gave it. The trainer thought the dolphin, being smart, might have remembered the order of the commands and simply done the set of responses from memory. So the trainer did what every good scientist would in those circumstances: she did the command gestures in a different order. "The dolphin still responded correctly before I gave the commands," she told Karen.

In fact, what the dolphin had been picking up was not the arm gestures, or the trainer's thoughts, but a slight shift of the trainer's body in a particular way before she moved her arm. Each arm gesture had a subtle body movement associated with it that preceded the gesture itself, and this was what the dolphin was detecting and responding to. Like Clever Hans, this dolphin had supersensitive visual perception (perhaps there was an acoustic cue as well) enabling it to pick up tiny body movements that the trainer was completely unaware of and that were imperceptible to most spectators.

When we think about dolphins being smart, we tend to think about intellect in the realms with which we are familiar, things that we humans do. These incidents of cuing tell us that these creatures are smart in ways that ordinarily we don't even consider. I will give animal-language critics credit for one major point, however: whether chimps, gorillas, parrots, or dolphins can be said to have any language ability, in the sense that humans use the term *language,* is a very complex question. We know they communicate, and we know they are intelligent. But should we expect them to have a sense of nouns and verbs? Or — in my case — should we let ourselves be open to whistles or to something complex yet fundamentally different from human syntax?

⌘

A few years after the conference, I spent the night at my lab and pondered some extraordinary sounds as I lay on a small cot in my lab, just feet from the dolphin pool. Hydrophones picked up whistles from the water as Terry, Circe, and Gordo felt the need to break their silence, and these somewhat eerie sounds were pumped into my room. Often, the three dolphins were quiet. Dolphins are most vocal when they are at a distance from one another, and when in close proximity, they apparently turn to other means of communication. At night, the distinctive guttural roar — *aaouu . . . aaouu . . . aaouu* — of nearby lions added to the acoustic tableau. All this took place against a constant background of chirping crickets at Marine World Africa U.S.A. in Redwood City.

The research pool at Marine World was in the middle of a raised wooded mound in what was effectively a six-hundred-acre wildlife theme park. Aside from the nearby lions, there were two water buffalo, Waldo and Wilma, close by, housed just behind me in a large enclosure. They looked fearsome, these giant bovines, but they were softies, really. My path to the pool each morning took me past them, and when I stopped at their fence they would come over to me, eager for the apples I always brought for them. I rubbed their huge, curved horns and patted their big black noses. They were adorable. Off to another side was a chimpanzee enclosure, from where we occasionally heard excited hoots when food arrived. An exaggerated mural of the African savanna covered a long wall nearby. So, although a good deal of excellent research was done at Marine World by local scientists, the area didn't give the appearance of the usual research environment.

One day soon after I began my work, I was by the pool feeding Circe. I happened to look up, and I saw a group of people by the wall, staring at me. That wasn't especially unusual; this pool wasn't accessible to the public, and I was used to people peering over to see what I was doing. But this morning was different. They weren't staring out of curiosity. They had looks of mounting horror on their faces, and some of them started to run. I looked at them, wondering what was so alarming, and then I felt a gentle nudge at my back. I turned, and there was Waldo staring at me, as friendly as could be. Someone must have left the gate to his enclosure open. I scratched Waldo's nose, got out my walkie-talkie, and said to security, "We have a Houdini" — the code many zoos use when an animal has escaped. I added, "Get me some apples." Waldo was very easily led back to his area.

❦

Ironically, Terry and Circe were thought to be "loser" dolphins, as both research subjects (according to none other than John Lilly) and performers (according to Marine World trainers). Lilly had heard about my program at Marine World, and he'd contacted the owner and negotiated his own use of two pools in an area behind the scenes. He set up shop and began a new but short-lived research project, Project Janus, trying again to train dolphins to speak English. He brought two female dolphins with him. One died due to a complicated pregnancy. The other, Terry, seemed too aggressive for the project and was transferred into our research pools. Terry actually turned out to be a sweetheart, and each time I arrived at the pool she greeted me by swimming rapidly around, porpoising occasionally, and then coming to where I stood and looking at me with big eager eyes. Terry was older than Circe, more mature, and she behaved like the matriarch. I worked gently with her to get her to trust us, and eventually she did. She was also very affectionate.

As for the younger female, Circe, the Marine World trainers had tried to work with her but described her as standoffish. She didn't like to be approached. Initially I had to keep my distance. I actually found this quite intriguing, and I thought of her as a kind of Asperger's syndrome dolphin. We gradually negotiated a relationship and eventually bonded very closely. The wall of the pool extended about three and a half feet above the ground, up to my waist. I could almost be face-to-face with Circe when she came to the edge. Many times we would gaze eye-to-eye for minutes on end, and I could sense her relaxing as we did this. Then she would roll over and present her belly to be stroked. When I did, her whitish skin would become rosy-colored, a sign of arousal. She liked me to rub her tongue, which seems to be a special source of pleasure for dolphins (Circe in France and other dolphins I've been with seem to like this). I have to admit that Circe was my favorite, and when she gave birth to Delphi, it was like my having a grandchild. My tight bond with Circe carried over to Delphi, and he became my muse. I loved these animals dearly, and I proudly insist that this does not get in the way of doing good science. In fact I think it helps.

And then there was Gordo, a Pacific bottlenose dolphin, a big male, eight hundred pounds, and the gentlest of creatures. I thought of him as a couch potato of sorts. In the spring of 1984, almost two years after I started at Marine World, Gordo's health began to decline rapidly and for no reason that the multitudes of vets that examined him could discover. He was very large. His problem may have been hormonal; his blood numbers looked bad. We rigged up a giant sling so that he would be supported in the water. He was weak for some days, and I could see he was slipping away from us. I was very worried and upset. I was in the pool with him often during these final days, usually along with a couple of trainers, giving him comfort, holding up his body when he was not in the sling. As I stood by him, I kept thinking of the ancient myths and modern news accounts of dolphins rescuing drowning people, holding them up, staying by, taking them safely to shore.

One evening, after I'd been in the water with Gordo for five hours, rubbing his skin to comfort him, the phone in my office rang. An assistant shouted to me that I really should take the call. I was reluctant to leave Gordo, but I did. As I was getting out of the pool to get to my office just fifteen feet away, I looked back, and I could see him straining to look in my direction. I made the call short, explaining the situation, but when I returned and placed my hands on him again, Gordo was gone. Dolphins can live to be over sixty years old, but the average lifespan in the wild is unknown. In my thirty years of research and rescue work, I have become close to many dolphins, of which a few have died. Every death is like a family member's passing; I mourn the loss of each deeply.

⌒∞⌒

The big question when I arrived at Marine World in the fall of 1982 was, Where will the money to build a research facility for a newly minted PhD who has almost no research experience come from? With the brash confidence of youth, I assumed the money would, as Gandhi once put it, "come from where it is now." I closed up my apartment in downtown Philadelphia and moved to an apartment in the Pacific Heights District of San Francisco. I landed a part-time teaching job in the Speech Department at San Francisco State University. A second part-time job doing biomedical research at Stanford Research International, in Menlo Park, boosted my tenuous financial situation, but it meant that I had to get to SRI by five in the morning, work until ten, and then go to Marine World to pursue the task of establishing a research facility. It was a tough schedule, and it didn't provide sufficient funds for building and equipping the kind of lab I had in mind.

I had an audacious idea. "Hello, I'm Diana Reiss," I said when Dr. Bernard Oliver answered the phone. "I just got my PhD. I'm starting a dolphin lab with a research program on communication here at Marine World, and I would love to talk to you about my research." Barney Oliver was founder and director of the Hewlett-Packard laboratories, and he'd played a key role in developing the HP-35 pocket calculator. He was also deeply involved in NASA's Search for Extraterrestrial Intelligence project, or SETI. That was my motive for contacting him. We had a lot in common, the SETI group and I, I reasoned. The people at SETI were trying to detect signals from an alien intelligence and find a way to decode them, and I was planning to record signals from a nonterrestrial intelligence and find a way to decode them. The only difference between us was the source of those signals. "Your work sounds very interesting, Dr. Reiss, but I am very busy right now," Barney said politely. In retrospect, I suppose I was obnoxiously pushy; I said, "Could I just come down and have lunch with you?" He was very kind and said, "Okay. How about tomorrow?"

What I didn't know at the time was that John Lilly had gone down this same path of reasoning more than twenty years earlier, and the SETI people had been very receptive to it. SETI was effectively established at the now famous 1961 meeting in Green Bank, West Virginia, at which the astronomer and astrophysicist Frank Drake introduced his famous equation calculating the number of civilizations in our galaxy that could potentially communicate with us. (It was known variously as the Drake equation, the Green Bank equation, and the Green Bank formula, and the solution is around ten thousand.) The meeting's attendees had come together to talk about the possibility of detecting intelligent life in the Milky Way, and they included astronomers, physicists, biologists, social scientists, and industry leaders. This group later came to be known as the Order of the Dolphin, in recognition of the common goal of its astrophysicists and biologists: to establish communication with "alien" intelligence. When I arrived on the scene, SETI was located in Mountain View, just thirteen miles southeast of Marine World. Very convenient.

The outcome of the lunch I had so audaciously pushed for with Barney Oliver was the evolution of a symbiotic relationship between me and SETI people, which included Frank Drake, Carl Sagan, and Jill Tarter, director of the SETI Institute and the woman on whom Jodie Foster's character in the film *Contact* was modeled. I soon found myself on discussion panels with eminent astronomers, astrophysicists, and other scientists so notable that I never would have imagined I'd even meet them, let alone sit side by side with them debating issues of mutual interest in an atmosphere of intellectual collegiality. I contributed articles to publications on exobiology, one of which was a paper with the title "The Dolphin: An Alien Intelligence."[3] The title was a bit of a mistake. It should have been "A Nonterrestrial Intelligence," but "Alien Intelligence," a phrase I used from time to time, mostly as a joke tailored for that community, slipped through. It didn't matter, because the issue was the same: How do you decode a signal that is completely foreign to the knowledge and experience of your species?

In addition to the intellectual relationship that was established that day at lunch, a more practical relationship began as well. Barney generously donated some of his personal HP stock, which allowed me to get my lab up and running and was sufficient to keep it operating for several years. Despite this, I was still on a very tight budget, but I was lucky with my subject of research; it made raising funds easier than it would have been if I were studying, say, the digestive system of liver flukes. I was frequently on the phone saying something like "Hello. I am doing research on dolphin communication here at Marine World, just down the road from you. We have very limited funds and I was hoping you might consider donating [some piece of equipment] in support of the research program." The answer was usually "Yes, willingly." For instance, I was given a generous supply of audiotapes from Ampex Corporation, a neighbor in Redwood City. With the bounteous support of Ampex, several other Silicon Valley companies, and the U.S. Geological Survey, I was now ready to embark on my research program proper.

⚬⚬⚬

Terry, Circe, Pan, and Delphi were the first dolphins with whom I tried keyboard studies. The experimental setup was fairly simple, too simple probably to allow the dolphins to show us what they were really capable of, but I was constrained by engineering issues. I was lucky to have a terrific engineering student, Bill Baldwin, who helped me construct my equipment, a twenty-one-by-twenty-four-inch underwater keyboard with nine positions on it in a three-by-three matrix.* Designing an underwater keyboard for dolphins is not a trivial engineering task, given the need for safety, sensitivity, and speed of switching in a corrosive, saltwater environment; water is a hazardous environment for most electronic equipment. Our solution was to power the keyboard by light, via fiber-optic cables, so that when a key was pressed by the dolphins, a light beam was broken behind the key. Since my lab was generously supported by Barney Oliver, the founder and director of Hewlett-Packard laboratories, he paved the way for his company to donate the fiber-optic cables.

When I began the study, I had only three labeled keys: a circle, which represented "fish"; a triangle, which represented "ball"; and an H, which represented "rub" — that is, one of us would rub the dolphin. Bill had cleverly engineered the keyboard system and software in such a way that if a dolphin touched, say, the triangle key, a specific computer-generated whistle would sound in an underwater speaker for the dolphins, and in my headphones I would hear the word *ball* produced by a computer-generated voice. My job would then be to give the dolphin the ball. The same thing happened for the fish key and the rub key. I would give a fish or a rub, depending on what I heard in my headphones.

In another laboratory I had heard examples of computer-generated sounds that were supposed to be akin to dolphin whistles, and for the most part, they sounded exactly that, computer-generated and tinny. I wanted something that at least sounded biological. A sound engineer from Mountain Music Systems in Silicon Valley, not far from the lab, helped me to create synthesized dolphin whistles. The computer whistles needed to be similar to natural dolphin whistles in frequency and time so they could easily be reproduced by the dolphins, yet they also needed to be different from the dolphins' own signals to show us that the dolphins could acquire new

---

* This experimental design was inspired by Sue Savage Rumbaugh and Duane Rumbaugh, who used keyboards to communicate with chimpanzees. For video of dolphins using keyboards, see www.hmhbooks.com/dolphinmirror#keyboard.

sounds. Through trial and error we found waveforms that generated sounds rich in harmonics that had the timbre of dolphin signals. But, of course, what sounded "dolphinish" to our ears might not have sounded the least bit so to dolphins. Could we create a shared code for humans and dolphins? Could we discover anything about dolphin learning?

Dolphins are adept at mimicking sounds. I therefore expected these dolphins to imitate our artificial whistles in some way, and I hoped to learn more about the process by which they learned these whistles. Dolphins produce a wide variety of whistle types and little was understood of their structure and function. I grandly hoped my keyboard could be a Rosetta stone for helping our decoding efforts.

I began the keyboard study on July 13, 1984, when Pan and Delphi were eleven months old, by simply lowering the keyboard into the pool and recording what happened. Terry and Circe were swimming with their respective offspring, as usual. Each dyad was in what I called a P2 swim formation — side by side within inches of each other. Eventually Terry and Pan approached the keyboard, apparently curious to investigate it. They got very close, and then Terry touched the triangle key with her rostrum. The ball whistle immediately sounded in the underwater speaker right next to them. But they turned and fled at high speed before I had a chance to give her the ball. They continued to swim rapidly around the pool for a while. This novel sound had obviously freaked them out, and they didn't come near the keyboard again during the rest of the half-hour session.

A few minutes after I put the keyboard in the pool the following day, Terry and Pan again approached it, this time much more cautiously. Then Terry pec-rubbed Pan, an act of reassurance, as if to say, Don't worry. It's okay. You try this time. He moved forward and touched a key, and it was like a light bulb went on for him; he quickly got into pressing the keys, hearing the whistles, and getting the objects or rubs. I had designed the protocol so that the positions of the keys changed every minute. The keyboard would be deactivated, and I'd hear in my headphones a mechanical voice saying, "Ball — position one; fish — position three; rub — position four," or whatever the preassigned pseudo-random set was. We changed the positions of the symbols from key to key at the end of each minute so the dolphins would have to learn which symbol was associated with what object or activity rather than erroneously thinking that the position of the key was the critical factor in obtaining what they wanted.

Delphi had been watching what Pan was up to with the keyboard, and two days after Pan's first tentative touch, he joined in too. (Pan was always first to do new things, and Delphi generally followed suit shortly thereafter.) Very soon we had the two boys enthusiastically using the keyboard, just as we had hoped. Terry and Circe, meanwhile, had no interest in it. Why, I'm not sure, but they seemed to be happy to use it as a baby-sitting device. Both mothers were constantly on the go with their calves, always attending closely to them, always swimming with them. Before the calves were born, Terry and Circe regularly rested or slept for ten-minute intervals. They would lie stationary at the surface or swim slowly with one eye closed; remember, dolphins are "specialized" sleepers. But once the boys were on the scene, there were no more breaks, which any human mother can understand. The only time Terry or Circe got to rest after that was when one of them took care of both calves for a while. This form of baby-sitting is called allomothering, and it is observed in both wild dolphins and dolphins in aquariums. Now that the keyboard was in the pool half an hour each day and the boys were enthusiastically using it, both mothers could get some rest. I empathized with them.

Both Pan and Delphi were still nursing at this point, as they would continue to do for another year or two, at least. Dolphins usually nurse for about three years, and sometimes longer, even though they are eating fish as well. Pan had already started to take an interest in fish; at first he'd simply mangled them and generally seemed to be playing at or practicing eating fish, but now he had started to actually eat them. It soon became clear that our inclusion of a fish key, with the promise of a real fish when it was touched, had been a mistake. In each session, Pan spent most of his time requesting and getting fish — small silver smelt — and used the other keys for ball and rub less frequently. (Rub was the least-favorite request throughout the study — perhaps, we thought, because they could get rubs from each other, while they depended on us for balls and fish.) We decided to remove the fish key after thirteen sessions and replace it with a symbol meaning "ring," a category of toy that the dolphins played with often.

At the beginning of the fourteenth session, Pan was the first to come to the keyboard, as usual. He normally pressed the fish key as soon as he got to the keyboard. This time he paused for a few seconds, scanning the keyboard. It was as if he were thinking, *Okay, where's the fish key?* He didn't try any of the other keys. Instead, he swam to the bottom of the pool, apparently looking for something. I had no idea what. He swam back to the keyboard with a silver smelt in his mouth, left over from the morning's feed. He held the fish up to a blank key and looked me in the eye expectantly. I could almost hear him asking for the fish. Both Pan and Delphi had hit blank keys a couple of times in the early sessions, but they'd quickly learned that those keys produced nothing, not a whistle, not an object, and they hadn't touched them since. The fact that Pan was using a blank key in the way he did was quite remarkable, an obviously deliberate act of attempted communication.

I was suddenly in an awful position. I knew in my heart and in my mind that he was trying to communicate with me in a very innovative way, and I would have loved to give him the fish. But that would have been against the rules of the game. The rules of the game were that he got objects when he touched the appropriate keys. And there was no fish key on the keyboard. I had no choice but to ignore him. He continued to look at me for a few seconds, as if he were thinking, *What don't you understand about my message? Can't you get what I'm asking for?* He soon gave up. A good example of scientific rigor working against fundamental decency and perhaps discovery. I *really* wanted to give him a fish.

⚬❈⚬

On the nineteenth occasion that Pan hit the ball key, a few weeks into the project, I heard something in my headphones that made me start. I heard the computer-generated ball whistle, of course. But there was something else. *Did Pan make something like a ball sound?* I thought to myself. I couldn't tell. As I said earlier, unlike dolphins, we humans are not superb acoustic detectors, especially in these circumstances. Pan was pushing his newly acquired ball around. He pushed it against the wall near the hydrophone to my right, and I heard another very clear ball signal, repeated twice. He came back to the keyboard, and this time he clearly produced the ball whistle and then pressed the ball key; again I heard something in my headphones that sounded a lot like Pan making the ball whistle. The same thing happened a third and fourth time. I couldn't wait to get to the lab to print out a spectrograph (a sound picture) of what I'd been hearing. Then I would know if my ears had been fooling me, if it had been nothing but wishful thinking.

What I saw set my heart pounding. It was one of those moments a scientist sometimes experiences, when you know you have seen something new, something that no one has ever seen before. It was a delicious moment. The ball whistle is quite simple: it's a sinusoidal waveform with flat ends, an initial rise, followed by a fall, then ending with a rise, all of which takes about a second. What the spectrograph showed me was that on the first occasion I thought I'd heard Pan make a ball sound, he had in fact produced just the end of the whistle, the final fall and rise, a U-shaped component with a flat tone at the end. On the second occasion he had made the beginning of the sound, a rise and fall. On the third occasion he had matched the timbre, the overall resonant quality of the model sound, by adding harmonics, and in his forth rendition, he'd brought it all together in a beautiful facsimile of the entire ball whistle. In each case, he matched the exact duration of the model sound, but he transposed the frequency a bit, whistling just a little higher. Pan had spontaneously mimicked a new sound, with no teaching, no training, no food reward. No reward at all, except what pleasure he might have derived from learning something new in an interactive, social environment. This was entirely novel.*

By repeating the last part, then the first part, then putting it all together, Pan had done what young children do when they learn new words. It's called segmentation. Kids often imitate the last part of a new word first, such as saying *nana* after hearing *banana*. In psychological terms, this is called the recency effect; it refers to the tendency to best recall what has been perceived most recently. Conversely, repeating the first part of a word is called the primacy effect. Then kids begin to put the two parts together.

When a child is able to put a whole word together, he or she often practices it. For instance, a child who has learned the word *dolly* can often be heard repeating to herself "Dolly, dolly, dolly" while playing with the doll. And this is what we saw with Pan and, subsequently, with Delphi too.

Both Pan and Delphi learned to produce the ball whistle after about nineteen or so exposures to the sound. They subsequently learned the rub sound, the ring sound, and the disk and float sounds that we later added, each one after fewer and fewer exposures. At first, Pan and Delphi imitated the sound immediately after hitting the key and hearing the computer-generated sound; we call this mimicry. But they made the sound at other times too, such as before pressing the key or later or while playing with the ball or being rubbed; we call this production. The difference between mimicry and production is that production is an indication that the dolphin

---

* Video of Pan first imitating the ball whistle can be found at www.hmhbooks.com/dolphinmirror#whistle.

has made a strong association between the sound and the object or activity it's associated with. When Pan or Delphi repeated the ball whistle when playing with a ball, it was the equivalent of the young girl saying "Dolly, dolly, dolly" when playing with the doll. Repetition serves to reinforce the word in the utterer's mind and perhaps forms associations in the dolphin's mind.

Whether these sounds represent to the dolphins what we might call a label, or even a word, is hard to say. And I was very cautious in my interpretation. I stressed that while the dolphins' productive and contextual use of their facsimiles clearly indicated that they had formed close learned associations among the sounds, visual symbols, and objects and activities, I couldn't conclude that the symbols or sounds represented words or labels to the dolphins. But it looks that way. Neither Pan nor Delphi ever whistled the sound for, say, "ball" and then hit the rub key or some other inappropriate key. They were 100 percent consistent in associating the appropriate sounds with the visual symbols.

On two occasions, Pan appeared to use the ball sound as a label while communicating with Delphi. The first time was at the end of the day. I wanted to clear the pool of toys, so I gave Pan the hand signal for "fetch." I thought there were a couple of balls left in the pool. Pan swam off looking for toys to bring to me, but there weren't any floating around. The only remaining toy was a ball that Delphi had in his mouth. Pan went head-to-head with Delphi. I heard the ball sound on my headset, Delphi released the ball, Pan took it in his mouth, and they both came to me at the side of the pool. Did Pan "ask" for the ball by whistling the ball sound? Did Delphi respond to the "request" by giving the ball to Pan? I can't say for sure, but that's what it looked like. These are the kinds of incidents that give you some real insight into how smart dolphins are. But because of the rigors of the scientific method, they are not suitable for reporting in a scientific paper. It happened on just a couple of occasions, and it was uncontrolled.

A year into the study, the results we were getting were exceeding my expectations. We had almost enough data to publish, but given my near-obsessive compulsion to get just that bit more in order to nail things down completely, not quite enough. I wanted more hours of observations, just to see what else Pan and Delphi might show us of their extraordinary minds. But suddenly the owner of Marine World, Mike Demetrios, announced that he was going to have to close the facility within the next few months and that all the animals had to be moved elsewhere. He'd found a new location for the park, in Vallejo, north of San Francisco, but there would be a two-year hiatus before it opened, from September 1985 to September 1987, and everything would be on hold until the new facilities were ready.

We moved the dolphins into their much larger pool complex at the new site in October 1985. All the animals — dolphins, elephants, tigers, otters, chimpanzees, water buffalo, sea lions, and many others — were safely moved into their new homes. But although the site was ready to house the animals, it was not yet ready to house humans. For two years, while the infrastructure and all the human facilities were built, we were unable to conduct research. The Marine World staff and our research team worked in temporary trailers on the fairgrounds across the road from the new site. Our team's primary concern was keeping the dolphins fed, healthy, and stimulated with enrichment activities and objects. The keyboard had to wait.

Finally, on September 28, 1987, we were able to begin again. I remember the date so precisely because as I began to lower the blank keyboard into the pool in preparation for a research session with Pan, Delphi, and their moms, we heard a beautiful production of the ball whistle. Who made it, Pan or Delphi, I wasn't sure. But here they were, not having heard the sound in the previous two years, and one had produced a near perfect rendition of the whistle. Had they been using it between themselves those past two years? Or did the sight of the keyboard elicit the response? Hard to say.

But this told us something about their minds and their memory for whistle sounds.

The observations of their behavior this second year, during which I was joined by Brenda McCowan, a doctoral student at Harvard, were definitely worth the enforced wait. Pan and Delphi now produced whistles much more frequently than they mimicked them. They often whistled the facsimiles when they interacted with the corresponding objects; they weren't just mimicking the sounds after pressing the keys. This increase in productive use over mimicry suggested a strengthening of the association between the sounds and the objects and implied that, for them, there was some social or functional use of the sounds. It's what you would expect in animals that are learning. The dolphins' own versions of the computer whistles showed great fidelity to the model sounds of their earliest imitations. But during the first and second year, we found that in some of their versions, the duration and frequency (pitch) of their renditions was altered, compressed or expanded. It was the equivalent of saying a word more quickly or more slowly, or varying the highness or lowness of one's voice. When we speak, we vary our voices. What the dolphins' variations meant to them, we don't know.

When we looked at the sonograms (sound pictures) of their calls, we could see that their versions of "ball" and "ring" and other computer-generated whistles were interspersed with whistles from their own repertoire. In this second year of the study, the dolphins' productive use of their facsimiles of the computer whistles increased significantly over their mimicry of the signals after using the keys. But what did this increase in productive use mean? Brenda and I conducted a detailed vocal and behavioral analysis of the video- and audiotapes recorded in all the keyboard sessions, trying to figure out the behavioral contexts in which they used their newly acquired whistles. We discovered that the dolphins were using the whistle facsimiles in behaviorally appropriate contexts! Appropriate contexts were narrowly defined as those in which a dolphin had physical contact with the specific object, not just when the dolphin was approaching or near the object. So when a facsimile of, for example, a ball was produced, it was in the context of ball play or contact with a ball. We know next to nothing about the overall repertoire of their calls, what each of them mean, and we know even less about the possible semantics or function of their more complex whistle sequences. But the fact that they showed such a proclivity to learn and use our artificial whistles correctly in context encourages the speculation that their natural whistles might also mean something; perhaps they are even references to objects (or individuals) and activities. We do know that among the rich and varied repertoire of whistles, squawks, and other types of calls the dolphins produce, the most frequently used signal is a contact call that conveys signature information about the caller. Dolphins use these relatively stereotypical calls and can produce the calls of relatives and other members of their social group. It seems that dolphins may call to or refer to other dolphins in what may function as a referential call.

In one study, Brenda and I found that the contact calls of female dolphins from the same social groups shared certain acoustic features.[4] I am now studying whether the acoustic features provide information about alliance, family, or social group membership. But as far as decoding the meaning or function of their whistles, we don't yet have our Rosetta stone. And although their whistles look structurally simple to us, this does not necessarily imply that they are simple to the dolphins or that their meanings are simple.

Pan and Delphi's biggest innovation in this second year of the study was the creation of a new type of toy play and the concurrent emergence of a new whistle. In this new type of play, Delphi would hit, for instance, the ball key, actively play with the ball for a few minutes, and then return to the keyboard and hit the ring key. He now had two toys. It isn't easy to hold an air-filled ball underwater, but he deftly managed it at the same time as he had the ring in his mouth, and then he and Pan would toss both back and forth. It took some coordination, but, as everyone knows, dolphins have excellent physical coordination. We called this new activity double toy play, showing how verbally inventive we scientists can be.

One day not long after the boys invented this new game, my colleague and friend Jill Tarter, the director of the Center for SETI Research, visited the lab. Jill spent a lot of her time looking at signals from space, and so she was happy to join me in looking over the many squiggly lines on the most recent spectrographs of the boys' whistles. She pointed to a particularly complex-looking whistle on one section on the spectrograph and said, "Look at that — doesn't that look a lot like the ring and ball whistles joined together?" I looked closely and realized she was right. I saw one long continuous whistle that began with a facsimile of the ring whistle and then connected to a facsimile of the ball whistle. This whistle was produced when the dolphins were engaged in double toy play with a ring and ball. During the following weeks, Brenda and I reviewed and analyzed the dolphins' vocal productions and discovered that they were indeed producing this novel whistle in the context of double toy play. They had invented their own new whistle with no guidance, reward, or acknowledgment from us; this was the dolphins' own doing. A reflection of these minds in the water.

I had designed the keyboard so that when one key was hit and the appropriate sound initiated, there was a brief refractory period, about half a second, during which there could be no second sound. This meant that Pan and Delphi had never heard the ball and ring sounds combined as a single, continuous whistle. And yet this was precisely what they were producing. By placing the two sounds together, were they making a statement about the spatial relationship of the ball and ring together? Or about the game, or about something we couldn't even imagine? I suspected dolphins created combination whistles with their own sounds in their own environment when they wanted to communicate information to other individuals.

When we wrote up our results for a paper,[5] we pointed out that previous attempts in other labs to obtain vocal mimicry in dolphins through active training had required as many as a thousand trials, compared with just a handful in our case. One of the goals of the study, we said, was to answer these questions: What do dolphins learn and how do they learn it when given the freedom to interact with a self-reinforcing system?[6] Pan and Delphi used that freedom to show us some of their best stuff. We had confirmed that dolphins join humans and some songbirds in the elite ranks of species that employ mimicry in vocal learning — and they do better on their own.

4

NONTERRESTRIAL THINKERS

JUST A FEW WEEKS after I began working at Marine World in Redwood City, a big package was delivered to the little research building I'd established next to the dolphin research pool. My brand-new Sony video camera had arrived, and I could hardly wait to try it out. "Let's go shoot some footage in the hydrolator," I said to Bruce Silverman, my research assistant. My intention was simply to try out the new gear and get the hang of it, not to capture something I'd never seen before.

The hydrolator was a very large, semicircular, glass-fronted elevator structure, positioned right next to one of the three large glass walls of the dolphin show pool. It was designed to glide slowly up and down, allowing people to get an underwater view of the action at different levels of the pool, which was more than twenty feet deep. But it never worked, so what we had was essentially a gigantic blue room. I had co-opted it to use as a private viewing station to observe Stormy and Schooner and the other dolphins that resided in the show pool. Bruce and I unpacked the camera equipment and did the little assembly that was required, and then I turned around to point the lens at the glass wall. There was Stormy right in front of me. Blowing bubbles. But they were no ordinary bubbles, and Stormy wasn't just idly letting air out of her blowhole. I watched as Stormy quite deliberately positioned herself horizontally almost on the floor of the pool and then moved her head upward with a small, quick jerk. At the same instant, she let out a large bubble that instantly formed itself into a perfect, lustrous bubble ring. It looked like a silver halo.[1]

I was transfixed. I had heard anecdotal accounts of dolphins and whales blowing bubble rings, but I had never before seen it. It was a magical moment. Before I had a chance to savor it fully, Stormy blew another ring. She looked up and watched as it slowly expanded and rose toward the surface. And then she blew another. This one seemed to be a little smaller than the previous one, and it traveled upward faster. Within seconds the new ring met with the other ring and, in a moment of undulating creativity, they coalesced to form a much larger, hoop-size ring. Stormy had been watching very carefully too, and as soon as the two rings became one, she quickly swam straight up and passed through the hoop, shattering the large ring into a cloud of small bubbles all around her. Bruce and I looked at each other, mouths agape with astonishment. We had just witnessed this no-handed dolphin create and use a toy using only her blowhole.

The infants of many nonhuman animals engage in play behavior with one another, of course. Puppies and kittens are at their cutest in rough-and-tumble play with their littermates or going over and around their ever-patient mothers. Most of us have seen video footage of chimp and gorilla infants frolicking. But this was different. This dolphin was actually creating an object of play and seemed to be quite aware of what she was doing. Stormy's behavior had every appearance of being premeditated and skillfully executed based on experience and practice, with nothing casual about it.*

Although a Rosetta stone to decode their whistles still eludes us, there are other ways we can glimpse the richness of dolphin intelligence. Some of their behavior (like ours) is genetically programmed, so there's no need for conscious thought. But much of the dolphins'

---

* For video of Stormy's bubble ring play and other dolphins producing bubble rings, see www.hmhbooks.com/dolphinmirror#bubble.

behavior, like that of other social mammals, is learned. A great deal of what they do is quite flexible and variable, based on their social and environmental circumstances. In this chapter we will consider three interactions of dolphins with their environment that give the impression of active minds at work and that demonstrate a degree of self-awareness that is rare in nature. The first is bubble-ring play and object manipulation; the second involves using the physical environment in innovative ways to forage for food; and the last is the sort of behavior that in humans would be called deception.

☙

Dolphins share an ancestry with even-toed ungulates, hoofed creatures such as bighorn sheep, goats, and cows, among others, although their last common ancestor dates back some forty-five million years. Odd as it may seem, given that length of time and their very different physical environments, dolphins and bighorn sheep share some body language. Both arch their backs as a sign of aggression; when surprised, curious, or frustrated, they both let out big breaths of air — this is called displacement behavior. When a sheep does this, it's not much to see. With a dolphin, of course, it produces a big, obvious bubble.

One of the first times I saw this in dolphins was in the early stages of the keyboard work, before I started any experiments. I needed to know how good the dolphins' visual discrimination was, both underwater and in air. Would flat two-dimensional symbols be best? Or should they be three-dimensional? I started off with flat symbols, and these seemed to work pretty well. But I thought I should test three-dimensional symbols too. The first time I did this, Circe took one look and let out a huge bubble, as if she were startled! These involuntary exhalations produce bubble bursts, which look exactly as you'd imagine.

Bubble caps are smaller, more sedate, and seem to be produced voluntarily. Bubble caps look like silvery mushroom caps that slowly rise through the water, swaying slightly and expanding as they ascend. Presley and Tab, two male dolphins that I worked with at a later period at the New York Aquarium, seemed to relish producing bubble caps, and they did it most days at a particular time. Dolphins normally feed underwater, of course. But Presley and Tab, like many dolphins in aquariums, were fed at the water surface, getting their fish as they held their heads above the water. After feeding, Presley went to his preferred area of the pool, and Tab went to his. The two young dolphins would then proceed, apparently quite deliberately, to release pockets of air from their mouths, producing a series of bubble caps, which they watched float shimmering to the surface. I speculated that they had taken in air along with the fish. It was like watching two gentlemen who, after dining at the club, repaired to the smoking room for brandy and cigars, and burped in appreciation of the fine meal they had just enjoyed. This was definitely Presley's and Tab's personal thing, as I haven't seen other dolphins do it.

While Presley's and Tab's postprandial display seemed to have a degree of intention and control to it, it doesn't match what is required for blowing bubble rings. When a dolphin releases a blast of air from its blowhole, the air bubble doesn't automatically form itself into a bubble ring; cigarette smokers cannot form smoke rings without a lot of practice. Although the physics of smoke being expelled into air is not exactly the same as air being expelled into water, a dolphin needs to exert the same degree of control over exhalation and tension of the blowhole to routinely blow perfect bubble rings. Dolphins usually have to practice, and they get better as they do, though I did once see Delphi blow a less-than-perfect bubble ring as a result of being surprised. (He quickly and enthusiastically perfected his technique.)

Dolphins may invent bubble play on their own and learn bubble-ring production from one another by careful observation, something we would call social learning. It's not surprising then that separate populations produce bubble rings by different methods.

This variation of style from one group to the next has all the hallmarks of the expression and transmission of culture. It's the product of active minds. I've already described Stormy's particular modus: the characteristic posture near the bottom of the deep pool, a quick upward jerk of the head with a concomitant release of a bubble that immediately forms a ring. Others in Stormy's pool followed very much the same procedure, although with some variations.

Dolphins are ardent contingency testers. On one occasion in this group, Stormy was blowing single bubble rings and watching them as they rose. She had four spectators, including Schooner, each intently monitoring the activity. Schooner suddenly peeled off, swam to the bottom of the pool, and returned with a half-eaten piece of fish. He deftly dropped the fish fragment in the center of the bubble ring, and it rose under the influence of the rising ring, spinning violently in the ring's vortex. The five dolphins watched with what to all appearances was rapt attention.

When the ring and its fish reached the surface, Stormy again swam to the bottom of the pool, took up the I'm-going-to-blow-a-bubble-ring position, and produced another perfect shimmering ring, once more watched by the other four dolphins. Schooner again left the group, retrieved a fragment of fish, and placed it in the center of the rising ring. The five dolphins watched keenly as the ring rose and the fish spun in its vortex.

These were active minds at work, testing contingencies: What will happen if I put a fish in the middle of the bubble ring? It gave me the surreal impression of a group of scientists doing experiments, making observations, testing the dynamics of a physical system. We weren't the only ones doing the experiments.

Dolphins can be very inventive in their bubble-ring play. At Sea Life Park Hawaii and at SeaWorld in Orlando, where the pools are relatively shallow, the dolphins turn to one side and blow rings horizontally. Then they spin the rings like kids with hoops, using deft flicks of their rostrums to spin the rings without touching them. Bubble rings tend to rise, of course, but the SeaWorld dolphins have perfected techniques for keeping rings close to the bottom. One trick is to keep the rings small. The dolphin does this by excising a small segment from a larger ring and then splicing the ring back together in what seems to be a single, expert movement of the rostrum. In a video sequence from SeaWorld taken by a spectator, one dolphin produced a ring, spun it a little, and deftly snipped off a smaller ring, which it then guided through the water. These dolphins and others at the National Aquarium in Baltimore also produce bubble rings and water snakes, undulating silver streams of air that they chase and then sometimes stitch together to form rings.

The main difference between Marine World and SeaWorld bubble-ring production is their orientation: vertical versus horizontal. But at Sea Life Park Hawaii, the dolphins have developed a third alternative, as Ken Marten and his colleagues described in a 1996 article in *Scientific American*. The dolphin creates a vortex in the water by a short, swift movement of its tail. The second step is injecting air into it. "The pressure inside a vortex is lowest in the center, or 'eye,' of the swirl," the authors wrote. "When the dolphin exhales into the vortex, the air migrates to the region of lowest pressure and is drawn out along the core of the ring-shaped vortex."[2]

One young female dolphin, Tinkerbell, was even more inventive. She released a string of bubbles while swimming in a curved path near the wall of the pool. "She then turns quickly, and as the dorsal fin on her back brushes past the bubbles, the vortex formed by the fin brings the bubbles together and coils them into a helix."[3] Tinkerbell had a second, similar technique in which she swam in a curved path across the tank, producing "an invisible dorsal fin vortex." She then retraced her path and injected a stream of air into the vortex, "producing a long helix that shoots out in front of her."[4] Tinkerbell was apparently a budding fluid dynamicist of some considerable skill, because none of the other dolphins at Sea Life Park matched these feats, although they did enthusiastically blow bubbles in more traditional ways, learning from one another, perfecting their techniques through practice, and always showing a lot of interest in other individuals' performances, just as we saw at Marine World.

In my current research program with dolphins at the National Aquarium in Baltimore, I've observed many of these same innovations in bubble play. Bubble play seemed to begin in this social group of dolphins with the introduction of Jade, a female dolphin from the SeaWorld "culture" of bubble-ring blowers. Her young offspring Foster (a male) and Bayley (a female), born at the National Aquarium, showed incredible creativity in their bubble creations in their first years. They often blew bubble caps and watched them as they rose to the surface. They chased water snakes they'd created by blowing air and then whooshing through it with their dorsal fins. When Bayley was two years old, she produced exquisite bubble rings with her tail, blowing air out quickly from her blowhole and flicking it with her tail in a movement that was so subtle we could hardly make it out even when we watched it in slow motion on the videotape. The young dolphins and some of the others at the aquarium also produced the classic vertical and horizontal bubble rings seen in other social groups.

Dolphins' bubble-ring play is very complex and demands at least a modest degree of skill in both producing bubble rings and playing with them. The dolphins *seem* to embark on the task with some planning and deliberation, they *seem* to know what they are doing, and they *seem* to have some grasp of the physics of how to produce them and how to manipulate them in creative ways. In other words, bubble-ring play appears to be the outcome of higher cognitive activities, the product of active minds. If different dolphin populations in different facilities followed the same methods in producing and playing with bubble rings or had only slight variations on a principal theme, then one could reasonably argue that it was the product of a basic dolphin behavior pattern and didn't require higher cognitive function. But this is absolutely not the case. There is tremendous variability in bubble-ring behaviors among different groups of dolphins, and within any one group, the patterns of behavior change over time. Active minds are at work.

∽∞∾

In the mid-1990s, a few years after I left Marine World in Vallejo and moved to New Haven, Connecticut, I had the chance to study bubble rings more systematically with Brenda McCowan.[5] She had completed her PhD and decided to continue the research in Vallejo as acting research director, working with a new population of dolphins, three adult females and a batch of youngsters. Brenda noticed that the dolphins had recently taken an interest in blowing bubble rings. She watched, casually at first, as infants seemed to learn bubble-ring production by observing their mothers doing it and then trying it themselves. They weren't very successful at the outset, but as they practiced, their expertise increased. Before long, the youngsters were quite proficient, and they became inventive. They enthusiastically manufactured bubble rings, developing ways to, among other things, spin the rings and flip them 180 degrees. What really caught Brenda's attention, however, was the youngsters' apparent enthusiasm for blowing double bubble rings, one right after the other, so that the two coalesced into one large ring, just as I had seen Stormy do many years before.

With our colleagues Lori Marino and Erik Vance, we collected data, systematically recording what the dolphins did, minute by minute, around bubble-ring play, capturing it on video and later having it analyzed by two different people.

Four youngsters, Avalon, Brisbee, Liberty, and Norman (aged between three and eight years), were participants in the study. After three months, the youngsters' enthusiasm for double-bubble-ring play began to wane, and they became interested in other activities. But in the meantime we learned several things. Although the dolphins were capable of blowing bubble rings anywhere in the water, from the bottom of the pool to near the surface, and although the quality of the bubble ring wasn't affected by where it was produced, the dolphins overwhelmingly chose to produce rings near the bottom of the pool. This strongly suggests that blowing a bubble ring is a planned event. Making the rings near the bottom of the pool allows more time and space for play, including producing a second ring to join with the first.

We found that the quality of the first bubble ring had a very large influence over whether a second one would catch and coalesce with it. An excellent ring (we rated them as excellent, good, fair, or poor) was twenty times more likely to have a second ring catch and coalesce with it than a ring of inferior quality. This meant that if the dolphins were indeed interested in two-ring events with coalescence, they had a strong incentive to monitor the quality of the first ring so they could decide whether to bother producing a second. That's precisely what we saw. Although a dolphin didn't always blow a second ring, the odds that it would do so after an excellent first ring were six times higher than after a ring of lesser quality. And the dolphins were always quick to show their displeasure at low-quality rings, instantly biting them or dispersing them with a flick of the tail. The dolphins monitored the rings with laserlike focus, watching what the rings did when propelled this way or that. This was a spectator sport for active watchers.

No matter what behaviors the dolphins got up to during bubble-ring play — whether one ring or two was produced, whether a ring was mediocre or excellent, whether the dolphin played extensively with the ring or not — each of them consistently brought the session to what seemed to be a formal close, biting and destroying the final ring before it reached the surface. Erik said that it seemed like "clearing the Etch A Sketch." Ending the sessions like that seemed to be as important to the dolphins as the sessions themselves.

These four young dolphins really did seem to know what they were doing, plan what they were doing, and respond to contingencies. It is remarkable what these animals do; they know something about physics.

Whenever I think about this study, an image always comes to me, something that Erik told me he saw toward the end of his first three-month stint: "After they had been fed, the dolphins liked to hang out and do bubble-ring play. That's what they did. One day while I was watching the four of them blowing bubble rings, I became aware of an infant beyond them, on the other side of a transparent barrier that separates the pool that houses very young dolphins with their mothers adjacent to the study pool. He was intently watching the older dolphins, almost with the awe that little kids watch older kids in the schoolyard. Then he spat out a few bubbles, and then a few more, trying to be like the older dolphins. He had a long way to go, this little infant. But he was trying."

⁓

Dolphins don't have hands, but that hasn't stopped them from being creative and manipulative in numerous ways. The principle in biology that function follows form is certainly reflected in the radical streamlining of the dolphin's body for life at sea. But forms can be co-opted for other functions as well. Stephen Jay Gould termed this exaptation.

The dorsal and pectoral fins of dolphins, like those of fish, function for stability, steering, and fine movements. (In dolphins, the dorsal fin and tail flukes are also involved in thermoregulation.) But dolphins use their fins for much more than hydrodynamics and locomotion. Dolphins will use one fin or another to touch, stroke, rub, caress, drag, slap, carry, and interact with other dolphins and a variety of objects. But even without using their fins, they can creatively manipulate their surroundings — as the bubble rings indicate. Here's another example.

In the early-morning hours at Marine World, after I arrived and before I headed to the research pools, I would generally pass the dolphin show pool just to check in on the other dolphins. I was often puzzled at the sight of a ring of river rocks encircling the central drain at the bottom of the pool. These smooth-surfaced eight- or nine-inch oval river rocks were strewn here and there on the pool bottom by divers so the dolphins could rub their bodies along them. The rocks were among a rich variety of toys and objects the dolphins received; there were also loofalike brushes and bumpy-surfaced mats attached to the pool walls for the dolphins' rubbing pleasure.

Yet overnight, mysterious rock circles were formed. A quick investigation revealed that the divers weren't responsible, so it had to be the dolphins' doing. How did the dolphins move these rocks? Rocks, like other objects, weigh less in water than out of it, so the problem wasn't the rocks' weight. The problem was technique. If they used their mouths, the dolphins would injure their teeth. Pushing with their snouts, the dolphins would injure their rostrums. But Stormy and the other dolphins had found a solution.

It took time, but finally we saw it. Stormy turned upside down, placed the top of her head on a rock, sucked it onto her blowhole, then turned right-side up and swam away with it atop her head. Ingenious! We saw some of the other dolphins carrying rocks this way as well. I never did determine why they were so intent on creating the rings. Perhaps it was a reflection of dolphin aesthetics.

⚬⚭⚬

Bottlenose dolphins live in a great variety of ecosystems, so it isn't surprising that they find different ways of catching prey. But dolphins are opportunistic feeders who use a large number of foraging techniques that are not driven simply by the characteristics of the different ecosystems. Different subpopulations learn different feeding strategies from their mothers. Dolphin inventiveness plays a part too, or so it seems.

One of the more bizarre foraging specializations in bottlenose dolphins is called sponging. Rachel Smolker and Andrew Richards first saw evidence of this innovative practice in September 1984. They were out in a boat in Shark Bay, in western Australia, during the first year of what turned out to be a very long-term study of bottlenose dolphins in the area. After they'd observed a group of foraging dolphins, they were surprised to see a single dolphin briefly surface. Smolker described the incident in her book *To Touch a Wild Dolphin.* "There was the big, lumpy, reddish brown blob that seemed to be attached to the front of the dolphin's beak," she wrote, "extending back toward its face." Smolker and her companion concluded that the dolphin had some kind of disease, perhaps a tumor. "This poor dolphin was suffering from some horrid affliction and still having to forage to ward off starvation," wrote Smolker. "I wondered if it was painful, if the other dolphins might avoid it."[6]

Soon the dolphin surfaced again. Smolker and Richards identified it by its characteristic fluke. But this time the "big, lumpy, reddish brown blob," the supposed tumor, was absent. It turned out that the "tumor" was actually a conical-shaped piece of basket sponge that grew in deep channels in the bay. The dolphin had apparently harvested the sponge and, for some reason, placed it on its snout. During subsequent years, the research team on Shark Bay grew. And that dolphin became the first of a small number of female dolphins studied for their use of basket sponges for foraging.

Bottlenose dolphins in the Caribbean sometimes forage for fish that are lurking just below the surface of the sandy seabed by using their sonar to locate the prey. Once the dolphin detects the prey, it launches itself into the sand, often burying its head up to its eyes and beyond. When the dolphin pulls out with the fish in its mouth, the sand falls back and forms a crater. After a group of dolphins has finished what is aptly called crater fishing, the seabed takes on the appearance of a moonscape. The elite group of females in Shark Bay forage in the seabed too, in deep-water channels. But they can't adopt the strategy of their fellows in the Caribbean for two reasons.

First, the sand in Shark Bay is extremely coarse and would scrape their beaks. Second: scorpion fish. These small, cryptic-colored fish are difficult to spot in the surface layer of seabed sand, and they deliver very nasty stings that any dolphin would want to avoid.

It is a challenge to make direct observations of dolphins using the sponges during foraging, because they are in deep, quite murky waters. And Shark Bay is called Shark Bay for good reason. Nevertheless, intrepid students, led by Georgetown University's Janet Mann, occasionally made research dives. They found that when they used the basket sponges as gloves on their hands, they could explore the seabed safely, ferreting out bottom-dwelling fish such as the spothead grubfish, which the dolphins in the area had been seen to eat.[7] Mann and her colleagues suspected that dolphins used their sponged beaks to pounce on the grubfish. There are a few very blurry photographs of dolphins apparently doing just this.

Sponging is a solitary pursuit, and those individual dolphins that do it spend far more time searching for food than those that use other methods. Nonetheless, the females that engage in it produce just as many offspring as females that have more time on their hands, so to speak, so the extra time burden is not at the cost of their reproductive success.[8] Mann and her colleagues believe that the solitary, time-consuming nature of sponging might explain why the practice is overwhelmingly done by females. Males, they say, are much too busy forming alliances with other males, plotting their future sexual conquests.

Sponging is a good example of tool use, an activity that is quite rare among nonhumans, mostly, but not exclusively, the domain of chimpanzees. Tool use in the oceans is even rarer than on land. Reports of tool use in a marine mammal such as the non-handed dolphin captured the attention of the media. When Mann and her colleagues published their scientific paper on the discovery of sponge-use by dolphins, reports of "first evidence of tool use in marine mammals" rippled through the popular press.

Genetic evidence suggests that sponging probably originated with a single female a handful of generations ago, a "sponging Eve." She passed it to her daughter through active teaching.[9] Sponging is therefore a very good example of social learning; one is tempted to refer to it as an example of dolphin culture.

Most foraging innovations by dolphins are highly cooperative. Mud-plume fishing, for example, or mud-ring fishing, as it is also called, was first seen in Florida Bay, in the Florida Keys, and necessarily takes place in shallow waters.[10] A group of around half a dozen dolphins patrols the waters in search of a school of fish. Once it is located, one of the group begins to swim in a large circle around the school, closing in ever more tightly. As the circle gets smaller, the ring-maker dolphin begins to beat its tail on the seabed, generating a plume of mud. The effect is a tightening net of muddy water that spooks individual fish to the point that they start to leap out of the water, where they are caught in the mouths of the waiting dolphins. Mud-ring fishing is very similar in principle to bubble-ring fishing in humpback whales, in which individuals swim in a circle below a school of fish emitting air from their blowholes, which creates a circular net of bubbles.

Dolphins (and whales) are masters at weaving variations on a theme, and so-called barrier fishing is a good example. It, too, involves trapping a school of fish by circling it, and then catching the fish as they try to escape. In barrier fishing, which has been seen off Cedar Key, Florida, the encircled school is driven toward a barrier of waiting dolphins. The trapped fish leap out of the circle, and many of them are promptly swallowed. A study of two groups of barrier-fishing dolphins in that area by Richard Connor and students at the University of Massachusetts at Dartmouth showed a division of labor between driver dolphins and barrier dolphins. In both groups there was just one individual that was consistently the driver dolphin, swimming circles around the fish.

Strand feeding is a cooperative endeavor as well, but it's very different from barrier fishing. I have observed this myself, as have many others, in muddy inlets in South Carolina and Georgia. It usually occurs within a few hours of low tide, when muddy banks are exposed. A small group of at least three dolphins patrols just offshore, looking for schools of fish, such as mullet. When they find one, they line up so the fish are between them and the muddy banks, and they begin to move cautiously toward the fish. Then, in unison, the dolphins surge to the shore, driving the fish up onto the banks. The dolphins follow the fish onto the banks, briefly beaching themselves, lie on their right sides to feast, and then wriggle back into the water. Rick Petricig, a graduate student at the University of Rhode Island, was the first to study this foraging strategy, which the dolphins apparently do both at night and during the day. It appears that the dolphins routinely strand themselves on the right side because the dolphin esophagus lies toward the left. Lying on their right sides, they can still swallow easily. Young dolphins learn this technique from their mothers, another example of social learning.

There are many, many ways that dolphins interact and cooperate. But their cooperation goes beyond their own species, and sometimes includes humans. I am not talking here about the swim-with-dolphins enterprises. I am talking about a form of cooperation that has gone on for at least two thousand years: a symbiotic relationship between dolphins and fishermen. It is a relationship in which both mammals benefit and learn from each other.

The earliest record of such an event was given by Pliny the Elder in the first century C.E. He wrote that fishermen would keep watch from the shore for passing shoals of fish that were too far out to reach from land, and if a pod of bottlenose dolphins happened to pass by at the same time as the fish, the fishermen would try to attract their attention by shouting, whistling, or slapping the water. Many years ago René-Guy Busnel, my professor in France, teamed up with the famed Jacques Cousteau to document this kind of activity in Mauritania. There, the fishermen didn't leave the arrival of the dolphins to chance; they engaged a shaman to pray and other-wise communicate with the dolphins so they would come close to the shore where the fishermen were waiting.[11]

Whatever brought the dolphins to the scene, their subsequent behavior was much the same wherever it occurred, according to independent reports over the centuries from the Mediterranean to North Africa to Australia: The dolphins moved toward the shore in unison, eventually trapping the shoal of fish against the beach. The fishermen used all kinds of means to collect their prey while the frantic fish leaped in all directions, and the dolphins grabbed their share by scooping them up from the air. It is barrier feeding, with humans attached.

However effective this system is, it pales in comparison with the choreographed operation that is initiated and controlled by dol-phins near the town of Laguna, in the southern tip of Brazil. Karen Pryor, who led the study, described the whole affair as "highly ritu-alized, and appears to involve learned behaviors in both men and dolphins." The fishing takes place on the shores of an inlet from the ocean near the center of town. The fishermen, armed with circu-lar throw nets rimmed with weights, line up in the water along the shore, or sometimes in boats. The dolphins patrol back and forth some fifteen to twenty feet offshore, facing seaward, looking for mullet. Pryor described what happened next:

"The dolphin reappears, usually in a few seconds, travelling toward the line of men. It comes to an abrupt halt and dives just out of net range, 5–7 m from the line, thus making a surging roll at the surface, a movement markedly different from normal respiratory surfacings. Men who are in front of the dolphin as it rolls then cast their nets ... The dolphins apparently take advantage of the confu-sion which the falling nets cause among the fish schools to catch fish for themselves."[12]

Pryor explained that the fishermen couldn't see the mullet in the water because it was extremely turbid. They observed the dolphins' behavior to know when and where to throw their nets. The men depended on the dolphins' cues, and "nets were not cast behind a dolphin, or in its general vicinity, but only in front of dolphins performing the correct behavioral sequence indicating the arrival of fish." The dolphins' rolls told the fishermen when to cast their nets, where to cast their nets, and, from the vigor of the movement, what size catch to expect. The entire arrangement is on a business footing: the fishermen catch the fish to sell, not for subsistence, and they do not attempt to feed the dolphins or touch them. Only about a tenth of the two hundred or so dolphins that live in and around the lagoon engage in this cooperative fishing. The fishermen recognize and have names for the ones who do and call them collectively the "good" dolphins.

According to town records, cooperative fishing has been practiced there since at least the mid-nineteenth century, with present-day fishermen saying that their fathers and grandfathers fished before them. And it seems very likely that the same familial pattern is true for the dolphins, because the practice is passed from mothers to offspring. (At the time of the study, in the late 1980s, one female, Chinelle, fished alongside two of her adult offspring and at least one grandchild.)

This last story of dolphins finding ingenious ways in which to manipulate their environment includes imitative behavior. Haig was a mature female bottlenose dolphin at the Oceanarium in Port Elizabeth, South Africa. She shared the pool with another mature female dolphin, Lady Dimple, and a Cape fur seal. The dolphins often imitated the swimming and other movements of the seal, but their imitations didn't stop with the seal. They also liked to watch as divers cleaned debris and seaweed from the bottom of the tank, Haig especially. Divers used a metal scraper attached to a suction hose; they scraped the seaweed from the bottom, and the loose debris was then sucked out of the tank. Both dolphins hovered in close attention, watching the divers' every move. Several days after the cleaning operation began, one diver accidentally left the apparatus in the pool overnight. When the diver arrived the next morning he found Haig busily at work trying to use the scraper and hose. She was "manipulating the apparatus by lying flat along the hose, which she clasped with her flippers, her rostrum resting on the metal scoop," wrote observers at the Oceanarium. "She investigated the apparatus from a variety of angles, manipulating it by pushing it in all directions and rolling it over." Although not exactly expert with the machine, Haig did manage to raise clouds of seaweed, which she promptly ate, something she routinely did in the pool even though bottlenose dolphins are not known to go down this dietary path in the wild.

Reluctantly, the diver took back the cleaning apparatus from Haig, who spent the next couple of hours roaming around the pool, apparently in search of something. She was then seen to be holding a broken piece of tile in her mouth, which she used effectively as a scraper "by swimming with the tile in contact with the bottom of the pool." She dropped the tile, ate the seaweed, and resumed scraping. Lady Dimple had been watching Haig during all this, and soon she found her own piece of tile, which she used just as Haig had. They were apparently excited by their new activity, as they were scraping off far more seaweed than they could eat. "The frequency of this behavior decreased with time," wrote the observers, "until the pieces of tiles were removed for fear that the dolphins would swallow them."[13] No doubt Haig and Lady Dimple went back to exploratory behaviors with whatever new resources they could find in the pool.

Sea otters also have an impressive variety of foraging practices, but they do not engage in the more cognitively demanding cooperative-feeding techniques. The degree of behavioral diversity in dolphins, including cooperative ventures, speaks of extensive behavioral flexibility rather than genetic hard-wiring. Dolphins are capable of such a large repertoire because they possess active minds. They can adapt, and they can innovate.

For dolphins, as for people, active social learning is key to individual and community success.

∽∞⌣

Deception is also a form of manipulation, but of other individuals rather than of the physical environment. Playing mental tricks on others and being the target of such tricks oneself is a common experience of human life. (It is sometimes called politics!) It is rare among animals. Three decades ago, Andrew Whiten and Richard Byrne at Scotland's University of St. Andrews surveyed fellow primatologists in search of instances of deception among apes. For one animal to be able to deceive another, it must have some idea of how its own behavior is seen by the other. To put it in lofty theoretical terms, the animal must possess some semblance of theory of mind — i.e., it must be able to put itself in the mind of the other. Deception, where it does occur, is likely to be a rare behavior, by its very nature; you can't cry wolf too often, or it won't work. Not surprisingly, Whiten and Byrne's survey turned up just a handful of credible instances of deception, and most of these were in the higher end of primate intelligence, among chimpanzees and a few large Old-World monkeys and baboons. They collected these instances in a book, *Machiavellian Intelligence: Social Expertise and the Evolution of Intellect in Monkeys, Apes, and Humans,* published in 1988.

But even when a good example seems to be at hand, it can be almost impossible to know how intentional the deception is. I'll give an incident that Whiten and Byrne saw themselves, which was actually what prompted them to survey their colleagues. In the early 1980s, Whiten and Byrne were observing the foraging behavior of chacma baboons in the Drakensberg Mountains in southern Africa. An adult female, Mel, was digging in the ground, trying to unearth a tasty tuber. Paul, a young male and a member of Mel's group, approached the busy baboon, looked around, and saw that there were no other baboons nearby. Suddenly, he let out a piercing yell, as if he were under attack. Paul's mother, who had been near but out of sight, came bounding over and chased Mel from the scene. When she saw that Paul was safe, she ambled off, probably to return to her previous activity. Now that he was alone, Paul walked over to where Mel had been digging and pulled out the bulb, which he calmly ate. Without Mel's previous excavation efforts, Paul would not have been able to dig up the bulb himself.[14]

On the face of it, Paul had just pulled off a successful deception. But did Paul really reason that if he let out a frightened yell, his mother would believe he was under attack, rush to his aid, and scare Mel away, thus leaving the bulb for him to eat? Or had he simply learned that if he yelled under these particular circumstances, he would end up with a bulb? It is impossible to say for sure. Yet overall, Whiten and Byrne's book contains enough credible examples of tactical deception to suggest that higher primates are capable of putting themselves in others' minds, turning what is usually an honest behavior into a dishonest one, and profiting from the experience.

As far as I am aware, no one has seen an example of tactical deception in dolphins and whales in wild populations. But I do know of examples of possible deception by dolphins in an aquarium setting in which the duped individuals are not other dolphins but human beings.

The first example, which achieved some notoriety in the dolphin-research world, occurred at Marine World in Redwood City a few years before I arrived there. Jim Mullen was chief trainer of the dolphins, and it was he who told me the story. Before he started working at Aquarama, Jim could be heard on street corners in South Philadelphia singing baritone with the 4 Epics, a rock 'n' roll group of the type that was common in the fifties and sixties. Despite a tantalizing recording contract with a famous New York company, commercial prospects for the group were not promising, and so Jim chose aquarium work instead, eventually becoming a very good dolphin trainer and an amiable colleague. (He also entertained us from time to time with some of his favorite oldies.)

One practice that Jim started soon after he joined Marine World in 1973 was to encourage the dolphins to tidy up their pool at the end of the day, training them with a reward of fish for each piece of litter they brought to him. "It worked very well," Jim told me. "The pool was kept neat and clean, and the dolphins seemed to enjoy the game." Then one day in the summer of 1978, Spock, always an enthusiastic litter collector, seemed to be especially diligent, taking piece after piece of brown paper to Jim. (Spock had been so named by an earlier trainer, an avid Trekkie.) Spock was rewarded with a fish each time he arrived with another piece of the brown paper. This continued for a while: more pieces of brown paper; more fish; very rewarded. Pretty soon Jim became suspicious, wondering where all these pieces of brown paper were coming from. He asked one of his assistants to go below and look through the pool windows to find out where Spock was getting the paper.

"It turned out that there was a brown paper bag lodged behind an inlet pipe," Jim told me. "Spock went to the paper bag, tore a piece off, and brought it to me. I then gave Spock a fish, as per our arrangement, and back he went. The second time my assistant saw Spock go to the paper bag, Spock pulled at it to remove a piece, but the whole bag came out. Spock promptly shoved the bag back into place, tore a small piece off, and brought it to me. He knew what he was doing, I'm sure. He completely had me." Spock's behavior looks very much like deliberate deception, because he hadn't been trained to take small pieces of litter to Jim and then be rewarded. He had been trained to collect any piece of litter and take that to Jim. By repeatedly tearing off small parts of the bag, Spock certainly maximized his reward. And when he pushed the bag back behind the pipe when it came out in one piece, that certainly had the ring of deliberate action. Whether you can call it deliberate *deception* is a tough call.

Here's a second example of possible deception, again at Redwood City, early in our exploration of keyboard skills. This was in 1984, when Delphi and Pan were just one year old. We had created arbitrary sounds with acoustic characteristics similar to dolphin whistles that would be specific for each dolphin, their whistle names, if you will. The way we did this was quite simple in structure, but it required each young dolphin to pay attention to the sounds we projected from an underwater loudspeaker, learn which was his sound, and then come to a specific station, his station, when he heard it. There was an underwater microphone, a hydrophone, near the speaker so we would be able to hear the dolphins' sounds too. If Delphi or Pan came to station when the other's sound had been played, I turned my back on him, like a time-out, to indicate that he had responded incorrectly. When the correct one came after we had played his sound, I rewarded him with a fish.

Delphi and Pan caught on very quickly, within a day or so each one coming to station only when his own sound had been played. About a week into this process, we were setting up for a session. I was at poolside, wearing headphones and carrying the control box from which I could activate the appropriate sounds. Bruce Silverman, my research assistant, also had headphones and a control box. We could talk to each other by intercom to coordinate our actions. I was just about ready to begin when I suddenly heard Delphi's sound in my headphones. I looked up and could see Delphi over by the speaker and hydrophone, eagerly on his way to his station. Bruce and I knew that I would initiate the session by using Delphi's sound key, but I hadn't touched it. *Hmm,* I thought, *Bruce must have forgotten and gone ahead and pressed Delphi's key.* I called to him and asked if that's what had happened. "No," he told me, "I haven't touched any key." I looked over at Delphi, who was by now at his station, mouth wide open, ready for a fish. I think Delphi had made his own sound over there by the speaker and then come for his reward.

Was Delphi trying to deceive us, trying to get a reward by manipulating the system? Not in the true sense, because we knew we hadn't pressed his sound key, but he had gone through all the motions as if we had. These animals are very smart at figuring out contingencies. In this case, Delphi may have thought something along the lines of *If I go over there by the speaker and my sound is made, then I can go to my station and get a fish.* He knew the rules of the game and had manipulated, or tested, them to his advantage. That's the basis of deception. And this time I gave him a fish!

Not long after the lab had been moved to the new site in Vallejo, Delphi seemed to have pulled another trick on me, and this time he really did fool me. In the wild, dolphins have to chase and catch their prey; we can't have them do that in an aquarium, for obvious practical reasons. But to make the feeding process as naturalistic as possible and hopefully more fun for the dolphins, we threw the fish out into the pool so the dolphins could chase them down. It worked very well, and Delphi in particular seemed to enjoy the game. One day my new lab manager and research assistant Laura Edenborough came to me and said, "Delphi is dropping fish all over the place. He's holding a lot of fish in his mouth, playing with them, and then dropping them. It's getting to be a husbandry problem, and there are a lot of fish lying on the bottom of the pool."

As I mentioned before, because Delphi was Circe's son, in some strange way I felt like he was my grandchild. I loved to feed him. So I sat by the edge of the pool, my booted feet on a little ledge. I indicated to Delphi that he should come over, which he did with his usual greeting display, and I played with him a bit. I turned to Laura and said, "All right, now I'm going to say to Delphi, 'Swallow it.' I'm going to give him a fish and say, 'Swallow it.' He doesn't know what 'Swallow it' means, but just use it as a mark or signal and he will learn to swallow the fish. When you see him swallow and look into his mouth and you don't see a fish, give him another one." I carefully went through this little regimen I had just described, and Delphi cooperated perfectly. He ate all the fish, and didn't drop any. Laura took over for the next several feeding sessions and all seemed back to normal — or so I thought.

After about a week with Laura and other students successfully feeding Delphi in this manner, I decided that I would do another feeding session. I stationed myself at the side of the pool and signaled for Delphi to come over. He did, and we spent a few minutes together. I rubbed him, and he volunteered his tail to be pulled, one of his favorite dolphin-human activities. Then I started giving him one fish at a time, waiting for him to swallow each before giving another. The session seemed to be going smoothly, but then I noticed that Delphi was making exaggerated swallows as I gave him each fish. When he opened his mouth after each swallow, there was no fish to be seen, so I gave him another one. Each fish was swallowed with a very big gulp. I had never seen him, or any of the dolphins, do that. I thought perhaps he had a sore throat that made swallowing uncomfortable. I gave him a couple more fish. More exaggerated gulps, more views of his empty mouth.

All of a sudden, Delphi's eyes got really big, the way they do when the dolphins are excited. Delphi opened his mouth, and I saw all these whole fish in there. He must have been holding them in his throat or had regurgitated them. Before I had time to open my mouth in surprise he started to shake his head, left and right, left and right. Fish flew everywhere. What a mess! Delphi was obviously having fun, and he had chosen to play this caching trick on me, not one of the students. I laughed hysterically; I couldn't help myself. Delphi had completely fooled me, completely manipulated me. And, from what I could tell from his demeanor, he seemed to know it.

5

THE FACE IN THE MIRROR

"WHAT ARE THEY DOING?" I exclaimed to no one in particular. I was perched on the observation deck overlooking one of the pools at Marine World, monitoring Delphi and Pan in the water below. The two young males, now seven years old, were engaged in sexual behavior, belly to belly. To put it technically, they were attempting intromission. In general, for bottlenose dolphins, copulation is the culmination of an aquatic courtship dance that I think they start to learn in the first few weeks of their lives — from their mothers.

I had been fortunate to be able to observe very closely the mother-infant interactions with Delphi and his mother, Circe, and with Pan and his mother, Terry, pretty much from day one. It was quite fascinating and revelatory — and it changed the way I viewed dolphin behavior from that time on. The behavior patterns that I saw with Terry and Pan were usually repeated within a few hours or a day later by Circe and Delphi. Whether that repetition was programmed, as Delphi was a day younger than Pan, or whether Circe was learning from the more experienced Terry, I can't really say.

In the first few weeks of life, calves, who don't yet have full coordination of swimming and breathing, swim right next to their mothers. This is called echelon swimming. In the mother's slipstream, the calf gets a bit of a free ride, expending less energy during the early weeks of life. Slipstreaming also keeps the calf in close proximity and tightly coordinated with its mother's movements.

Newborn dolphins are born without much blubber — a layer of insulation made up of fat and fiber, riddled with blood vessels, that helps dolphins and whales thermoregulate in cold and warm waters. This fatty layer provides a storehouse of energy for these mammals for foraging, migrating, and breeding. It also helps make them buoyant in water. The blubber layer is built up as the calf nurses on its mother's milk, which is 10 to 20 percent fat and rich in calories.

As the calf gradually gains better control over its breathing, it begins to swim just underneath the mother. We call this the baby position, and it has two big benefits. It facilitates nursing from the mammary slits that flank both sides of the genital slit, at the narrowing end of the mother's underside where belly meets tail. Also, it may confer some camouflage from potential predators below. In the vulnerable first weeks of life, the light underside of the calf may be less visible when it's silhouetted against its mother's light underside.

During these first weeks the babies begin to get bolder, darting off on little excursions of their own, forgetting that they haven't yet learned how to stop or return to their moms. The mother quickly swims after her calf and reestablishes the echelon or baby position. In the following weeks the calf learns to stop and return to its mom, and to swim in circles around her. Dolphins are social learners, observing, listening, and imitating what they see and hear. Whether adult dolphins or whales intentionally teach specific skills to their young remains unclear. Hal Whitehead and Luke Rendell have described what indeed seemed to be a case of such pedagogy in orcas, killer whales, who appeared to instruct younger orcas in predatory methods and maneuvers.[1] Terry and Circe, too, exhibited several instances of what could be loosely interpreted as instruction during the early mother-and-calf interactions.

For example, I observed Terry stop as Pan continued to move forward past her head, whereupon she pushed him backward with her rostrum, and positioned him perpendicular to her in what looked like a T formation. She then nuzzled his genitals with her beak. Terry patiently repeated this behavior with Pan: stopping, allowing Pan to go forward by himself, and then stopping him with her rostrum. She repeated this many times until Pan got the hang of putting his aquatic brakes on. A day later I saw Circe do the same sequence with Delphi. The calves began to stop on their own in the T position, lying on their sides, completely still. I called this the dead man's float (another of my highly inventive technical terms). As a reward, they received genital nuzzling from their mothers.

Very soon the babies became even more adventurous, embarking on more extensive forays on their own, which led to a game of chase; the mothers would swim after the calves, corral them, and bring them back. Before long, the babies learned how to circle back on their own, and the chase developed into what I called a figure-eight swim, which is exactly what it sounds like. The big adult and the diminutive calf chased each other in a figure-eight path, occasionally porpoising, or arching clear of the water in repeated jumps. During the chase the pursuing mother would sometimes approach her calf from behind and nuzzle his genital area with her beak.

Next, Delphi and Pan, in their initial and continued forays with each other during their first six months of life, did the same tango sequence with each other — the chase, the figure-eight swim, the dead man's float, the beak-genital nuzzling. Eventually, it culminated in what looked like sexual behavior. One of the boys would approach the other from behind and below and attempt to copulate with him. Dolphins copulate belly to belly. The boys would often switch roles, one pursuing, the other pursued.

The light bulb went on in my head. From the standpoint of dolphin social behavior, this pattern matched the components of dolphins' courtship and reproductive behavior. The male chases the female; they swim in figure eights; she floats on her side; he nuzzles and then enters her. In short, young bottlenose dolphins appear to learn this socio-sexual behavior through interactions with their moms. And in the case of Delphi and Pan, it was natural that they would interact in the same way with each other.

Growing up in close intimacy, Delphi and Pan were buddies.

They had a strong alliance, interacting and playing together in many ways. During the summer months, when it got hot in the Bay Area, things got hot in the pool at Marine World too. Delphi and Pan engaged in these sexual behaviors quite frequently, and in every part of their pool. Bottlenose dolphins show among the highest incidence of same-sex sexual behavior of any animal. Bonobo chimpanzees are much more famous for it, but dolphins match them in this department. Same-sex sexual activity is very common in nature, much more than most people realize.[2]

In both dolphins and bonobos, sexual activity of this sort provides behavioral glue in interpersonal relationships; in larger groups, it provides intergroup cohesion. Male dolphins form long-term alliances and work together cooperatively in many activities, including foraging and mating. This sociosexual behavior between male dolphins helps them form and maintain bonds; when performed with receptive female dolphins in estrus, it functions as reproductive behavior. Delphi and Pan's display of what looked like sexual behavior was actually social behavior — they may have been practicing, but they were also growing close.

Long before they reached age seven, Delphi and Pan had engaged in this chasing, soliciting, and intromission attempts. What caused me to exclaim out loud that summer day in 1990 was something I had not seen before. Delphi and Pan had very deliberately positioned themselves in front of a large mirror we had recently installed in the pool. It seemed to me they were clearly and intently watching themselves in flagrante delicto.

✺

We installed the mirror in the pool as part of a very serious scientific experiment. While I hoped the dolphins would like the mirror and find it engaging, the real reason for the mirror was an experiment simple in concept yet profound in implications for our understanding of dolphin mind and human mind. We wanted to find out if Delphi and Pan would recognize that what they saw in the mirror was a reflection of themselves, not just other dolphins. If they could, it would tell us that bottlenose dolphins possessed a level of self-awareness that is rare in nature. Not so very long ago, it was assumed to be the sole province of *Homo sapiens.*

Self-awareness — the capacity to have a concept of self and to know that one exists as an individual being — was long seen as a cognitive Rubicon: humans in supreme isolation on one side, specially endowed, and the rest of "base nature" on the other, possessed of brains of various sizes but with an absence of minds. It's a tempting thought for those who view *Homo sapiens* as the pinnacle of biological evolution, with cognitive competences far beyond and distinct from any other animal's. Conversely, for those who hold the view that humans are just one of many branches of evolution, with extraordinarily developed minds but not divinely ordained uniqueness, it is less shocking to find overlaps with other types of minds.

In 1970, Gordon Gallup Jr., then at Tulane University in New Orleans, upset the comforting, self-serving picture of human uniqueness when he demonstrated that the common chimpanzee, *Pan troglodytes,* also possesses a degree of self-awareness.[3] Gallup had shown that chimpanzees can recognize their images in a mirror as themselves. Gallup, who had long been a skeptic of "animal mind," was inspired to perform the mirror experiment by the account of a visit by Charles Darwin to the London Zoo in late March 1838. Jenny, a newly arrived orangutan (the zoo's first), was attracting a lot of public attention. Darwin wanted to see for himself. A few days later, he wrote the following in a letter to his sister, Susan Elizabeth Darwin:

"The keeper showed [Jenny] an apple, but would not give it her, whereupon she threw herself on her back, kicked & cried, precisely like a naughty child. — She then looked very sulky & after two or three fits of pashion [sic], the keeper said, 'Jenny if you will stop bawling & be a good girl, I will give you the apple.' — She certainly understood every word of his, &, though like a child, she had great work to stop whining, she at last succeeded, & then got the apple, with which she jumped into an arm chair & began eating it, with the most contented countenance imaginable."[4]

Darwin was impressed by Jenny's humanlike behavior, writing in his notebook of the day, "Man in his arrogance thinks himself a great work, worthy the interposition of a deity. More humble and I believe true to consider him created from animals." Darwin visited Jenny at the zoo two more times and wrote that she appeared to be "astonished beyond measure" when she saw her reflection in a mirror. Gallup recognized the significance of this little anecdote, and he devised what has since become known as the mirror test. Whether Jenny had indeed recognized her own reflection in the mirror that day, Gallup determined to find out if chimpanzees possessed the intellectual capacity to do so.

When an animal first sees itself in a mirror, it usually explores the mirror itself, touching it, looking behind it and over it. Then it typically behaves as if the reflection is another individual and it tries to engage it in social behavior of one kind or another, sometimes being aggressive, sometimes trying to elicit play. Gallup put a mirror in front of each of four chimps, two preadolescent females and two males, who all did exactly this at first. Very soon, though, the chimps started to engage in another type of interaction with the image, what is called contingency testing. Each chimp appeared to realize that what it did had some influence on what the chimp in the mirror did. It bobbed up and down, turned its head from side to side, that kind of thing, all the time intently monitoring what the "other chimp" was doing.

Relatively quickly, within three days, each of Gallup's chimps seemed to figure out that the chimp in the mirror was in fact itself, and they started to use the mirror as a tool. They would "[groom] parts of the body which would otherwise be visually inaccessible without the mirror, picking bits of food from between the teeth while watching the mirror image . . . making faces at the mirror, blowing bubbles, and manipulating food wads with the lips by watching the reflection."[5] From this point on, each chimp stopped all attempts at social interaction with the image in the mirror.

To any reasonable person this self-directed behavior, using the mirror as a tool to inspect themselves, is compelling evidence that the chimps recognized the reflections as themselves and thus were self-aware. But as a scientist, rather than just a "reasonable person," Gallup felt he needed to go one step further to prove beyond doubt that the chimps' behavior was what it seemed. An ultraskeptic could always argue, for instance, that the chimps' apparently self-directed behaviors were simply sophisticated forms of interaction with the chimp in the mirror. So, as Gallup wrote in his paper in *Science*, "In an attempt to add direct experimental support to the idea of self-recognition of the reflected image," he carried out what came to be called the mark test.

Gallup anesthetized the chimps and used a simple, odorless red dye to mark two spots on each: the uppermost portion of an eyebrow ridge and the top half of the opposite ear. After each chimp regained consciousness, Gallup observed its behavior for thirty minutes and noted how often, if at all, it touched the marked areas. Then he positioned the mirror close to the chimp's cage and watched its behavior. While the chimps had paid no attention to the marked areas before the mirror was reintroduced, they were now deeply curious, repeatedly touching the marked spots, sometimes inspecting their fingers, sometimes smelling their fingers after they had touched the marks. Clearly, they knew who was who. Gallup claimed that this was the "first experimental demonstration of a self-concept in a subhuman form."

By way of contrast, Gallup put three species of monkey through these same experimental hoops. None of them showed any indication that it thought the mark on the animal in the mirror had anything to do with itself. Moreover, unlike the chimps, the monkeys never stopped trying to interact socially with the image in the mirror, continuing to think it was another individual. "Our data suggest we may have found a qualitative psychological difference among primates," Gallup wrote in his now classic paper, "and that the capacity for self-recognition may not extend below man and the great apes."[6]

We had introduced the mirror into the dolphin pool at Marine World in Vallejo in July 1990 to test whether mirror self-recognition (MSR) was indeed the sole province of large-brained primates. In the two decades since Gallup had published his original study, orangutans, bonobos, and, to a less dramatic degree, gorillas had also demonstrated mirror self-recognition. Old- and New-World monkeys continued to fail the mirror test. In the evolutionary lines closest to mankind, a threshold had been drawn.

Why not bottlenose dolphins? After all, we knew by then that dolphins showed comparable abilities with the apes on other cognitive tests. And you don't have to be in close proximity to dolphins for very long to get the strong impression that they are indeed self-aware. Despite my scientific intuition on that matter, the notion that dolphins could exhibit MSR ran counter to the accepted idea of a single mountain of "higher intelligence," with humans at the summit. But the origin of human intelligence — including the capacity for spoken language and higher levels of consciousness — was typically cast as an enhancement of capacities that existed only in the common ancestor of humans and the great apes. From this "continuity" perspective, the intellectual capacities of *Homo sapiens* were viewed as superior to those of our closest primate cousins in quantity but not in kind. Such thinking was Darwinian — none of these scientists argued for divine intervention — but it was self-centered. Evolution was seen as a scaffold for humans, building on the kind of mental machinery that also exists in chimpanzees, gorillas, and orangutans and extending it to the levels we humans enjoy today. The continuity perspective is a primate-centered lens to explain why humans have the cognitive skills we do. It posits a cognitive gap between humans and great apes, and then between great apes and the rest of the animal world.

My intuition that Delphi and Pan were indeed self-aware was, of course, not acceptable as scientific evidence. The boys had to pass the mirror test. And to do that, they had to pass the mark test. We faced one blindingly obvious problem with the test: unlike primates, dolphins have no hands with which to touch foreign marks on their bodies. We used a highly technical term for this: the no-hands problem. As simple as it is, the mirror test is a much greater challenge for a species whose members lack hands (or trunks, or some such appendage) with which to touch their bodies.

A study of mirror self-recognition hadn't been on my original research agenda at Marine World. A few years earlier, Danny Povinelli, a colleague and later director of the Cognitive Evolution Center at the University of Louisiana, had called me and suggested the idea, and although I thought the dolphins would be excellent candidates for a test of MSR, at that point it just wasn't the right time, for me or the dolphins. We were newly into the keyboard work, and it was a very busy and sensitive period at the lab, so I opted not to pursue the collaboration then.

A year or so later I got a phone call from Lori Marino. She explained that she was a doctoral student in Gordon Gallup's lab, now in Albany, New York, and that they knew of my research and wanted to collaborate on a study of mirror self-recognition in dolphins. I didn't personally know Gallup yet, but of course I was quite aware of his groundbreaking work and greatly admired it. I was also intrigued by several of his ideas. Foremost among them was that MSR and empathy might be linked. For example, children begin to recognize themselves in the mirror at about the same time that they start showing concern or empathy for others, sometime between eighteen and twenty-four months old. Apes also show both empathy and MSR abilities, whereas monkeys show neither. This led Gallup to speculate that other species that showed social complexity and empathy, like dolphins and elephants, would likely have the capacity for MSR.[7] This time the timing was right, and I decided to do the study with them. Gordon, Lori, and I agreed to team up and test these minds in the water.[8] Lori came out to the lab shortly afterward, and we spent time pondering the no-hands problem and thought about how to conduct the study. We were in uncharted and very challenging territory.

So, when we introduced the mirror into the dolphin pool at Marine World in Vallejo in July 1990 to investigate Delphi's and Pan's responses to their reflected images, we challenged the primate-centered explanation of the origin of higher levels of self-awareness. After all, humans and dolphins have been on divergent evolutionary paths for ninety-five million years. If we were able to demonstrate that Delphi and Pan had the self-awareness to recognize images in the mirror as themselves, it would challenge the nature of human self-awareness. It could no longer be viewed as the culmination of smarter and smarter primate brains; self-awareness could arise in brains that were very different from ours in architecture and circumstances of development. Perhaps consciousness could arise in nature via many potential paths.

There were two circular pools at Marine World, each fifty feet in diameter and sixteen feet deep. They were joined by a channel twenty feet long, ten feet wide, and eight feet deep. We rigged up a PVC frame at the exit to one of the pools into which we could drop a three-by-five-foot two-way mirror, which we could cover with a black rubber tarp. We stationed an underwater video camera behind the mirror, in the interconnecting channel. During my early experience as a set designer, I had learned how to construct scenery from whatever was available and on a very limited budget. This experience as a bricoleur has served me well.

Underwater windows in aquariums usually have some degree of reflectivity, like poor mirrors. But the pools at Vallejo had no such windows, and so for the first seven years of their lives, Delphi and Pan had had no experience with their own reflections. We were dealing with completely mirror-naive test subjects. Our immediate task was to see how they would behave when they first saw their reflections.

Dolphins are curious creatures, always ready to investigate novel objects in their environment. So it wasn't surprising that when we lifted the tarp from the mirror, Delphi and Pan were eager to check it out. They touched its surface with their beaks and aimed a lot of echolocation signals at it. It must have been very odd for them to see what appeared to be three-dimensional objects in front of them, their reflections in the mirror, while getting a sonar signal of an essentially flat surface. But if they were surprised, it wasn't obvious to me. They quickly began to interact with the reflections in the mirror as if they were other dolphins. Very soon, though, they moved into what had every appearance of contingency testing. While staring at themselves, they circled and cocked their heads, rocked their bodies back and forth, and opened and closed their mouths. Delphi and Pan definitely appeared to be testing the consequences of their actions, just as Gallup's chimps had done. So far, so good.

On the eighth day of one-hour-per-day exposure to the mirror, both Delphi and Pan moved to the next level. They appeared to engage in self-directed behavior. For instance, they each got very close to the mirror, putting an eye just a few inches away, staring intently. They opened their mouths very wide, as if examining the inside, and then wiggled their tongues, something chimps did with mirrors. At different times, Delphi and Pan each positioned himself so he could see his own ventral surface and then protruded his penis, an action that is under voluntary control in dolphins. And of course there was the courtship dance and reciprocal intromission attempts with each other that I described earlier.

It was all suggestive evidence of self-recognition. But a skeptic could argue that the penis display, and even the intromission attempts, might be a sexual demonstration for the "other" dolphin in the mirror. The skeptic could also argue that the open-mouth behaviors were really open-mouth threats, something dolphins do during aggressive encounters. But this open-mouth behavior was very different; their mouths were relaxed and then held open in a prolonged manner. Also, the penis display had occurred when no other social behavior was directed toward the mirror. As for the intromission attempts, we noticed that when the mirror was covered and unavailable, Delphi and Pan did it in many different locations in the pool, but when the mirror was in plain view, they did it *only* in front of the mirror, always intently gazing at the reflection. And tongue wiggling, as far as we knew, was not part of any aggressive behavior.

On more than one occasion, Pan spent minutes in front of the mirror bending his head toward it, which gave him a view of his blowhole, and then he produced bubbles and bubble streams along with a variety of sounds. We had seen dolphins blow bubbles before, of course, but Pan was obviously looking at himself in the mirror. Delphi and Pan had pretty clearly gone down the cognitive path pioneered by Gallup's chimps two decades earlier. It was time for the crucial mark (with no hands) test.*

Our plan was that during a feeding session, we would put a mark on each dolphin's body in a place he could not ordinarily see. We would then give the dolphin a release signal to let him know he could swim away and do as he pleased. We marked one animal at a time. If the dolphin was curious about what had happened during the marking session, we expected he might go to the mirror and visually inspect the mark. Or, if he wasn't curious, we thought he might notice the mark in the mirror while swimming freely and then use the mirror to take a closer look.

The first session did not go well. Delphi was the first to be marked. Lori and I and my research assistants were filming and collecting data from an elevated observation deck twelve feet in the air, adjacent to the pool, and I asked one of the dolphin trainers

---

* For video of dolphin mirror play, see www.hmhbooks.com/dolphinmirror#mirror.

to mark Delphi. The trainer inadvertently put far too much of the mark (white zinc oxide, the main ingredient in many sunscreens) on Delphi's side. It was a huge, thick smear, and it appeared to really freak Delphi out. He broke station without waiting for the release signal and started speed swimming around the pool. I was upset that he seemed so upset, so after a few minutes I shouted to the trainer, "We've got to call him back to station, now! Forget the experiment. We have to get this stuff off him." My own heart was pounding as the trainer placed the bucket of fish at the pool's edge and put her hand in the water, part of the signal to come. Delphi came right over. The trainer gave him the hand signal to lay out, which the dolphins had learned to do for physical exams.

The trainer quickly wiped away the big white mark, and Delphi immediately relaxed. Again without a signal, he broke station and made a beeline for the mirror, orienting that part of his body where the mark had been toward the mirror. He appeared to inspect it closely. He then came back to station. We were thrilled. In that little episode, Delphi gave every indication that he knew the dolphin in the mirror was in fact him, and he had used the mirror to check out something on his body. We seemed to be on the way to that final crucial step in the test.

Unfortunately, the use of the mirror to inspect a part of the body *after* the mark was removed was pretty much the *only* such compelling behavior during the mark (with no hands) test. We conducted a few other mark tests, putting a less freaky amount of zinc oxide on both Delphi and Pan, but neither of them made a beeline for the mirror to check it out as we had expected them to. Only after we removed the marks did they unequivocally race to the mirror and quite deliberately orient their bodies to inspect the area. "Whereas the dolphins did seem interested in using the mirror to examine their bodies after the mark had been removed," we wrote later, "we did not find any other instances of posturing that was unambiguous." We were therefore forced to conclude that the work had been *suggestive* but not *conclusive* of self-recognition. We felt we had been so close. (I had conversations about this work with Ken Marten, a friend and colleague at Sea Life Park Hawaii, shortly after we finished this experiment. He decided to embark on a similar project and produced similarly inconclusive results.[9]) It was very frustrating. In retrospect, I think we probably stopped the work prematurely, partly through force of circumstances: I was pregnant and was planning to spend a year back east at Yale with my husband before returning to Vallejo.

❧

In his original chimp mirror self-recognition paper, Gallup had essentially said that the self-directed behavior he'd observed in the chimps he had tested was enough to convince a reasonable person that these animals were self-aware but he felt he needed a "scientifically objective" measure: the mark test. We were now in the same position with the dolphins: we all felt in our guts that dolphins, too, were self-aware, given what we'd seen, but they had just failed Gallup's litmus test. Why?

As I mentioned earlier, human children generally develop self-awareness between the ages of eighteen months and two years. This cognitive capacity resides in the prefrontal cortex and elsewhere and is associated with the ability to show concern for others. "Poor Mommy," a two-year-old might say if Mommy gets a boo-boo. But psychologists know very well that not all of them can do this by age two, not even neurologically normal kids, despite the fact that from a very young age all children are cued almost daily, as in "Hey, Morgan, that's you in the mirror. Don't you look *cute!*" When Morgan sees herself in the mirror when she's a little older, she already knows it's her. So when we insist that in order to demonstrate that a non-human animal is self-aware it must pass the mirror test, including the mark test, with no prior cuing, we are actually setting the bar far higher than we do for human children.

Recognizing oneself in the mirror seems like a simple act. You roll out of bed in the morning, you go to the mirror, you're either happy or not so happy with what you see, but you know without any effort what you are seeing, and you make use of the mirror to remove that eyelash resting on your cheek or to iron out that wrinkle inflicted by the pillow. Sounds simple, but in fact, cognitively, it is quite complex. In the mirror test, an animal must first pay selective attention to the information in the mirror. Many animals don't do this. Second, if the animal does pay attention, it has to interpret what it sees. Most animals that do pay attention to the mirror interpret the image as another member of their own species and try to engage it in some form of social interaction. If the rare observing animal recognizes the image as "self," it must then be motivated to use the mirror as a tool to observe and inspect itself before we can be sure what it knows. So, self-directed behavior requires both the *cognitive capacity* that underlies the concept of self and the *motivation* to use the mirror as a tool. Passing the mark test is yet more specific: it requires motivation or interest in touching the mark. I see that as a distinct barrier.

Chimps spend a good deal of their time grooming themselves and even more time grooming one another; dolphins, for obvious reasons, do not. Dolphins engage in high degrees of tactile and physical contact with others, both dolphins and humans, and their skin is highly innervated so they are sensitive to touch. But for the most part, paying attention to marks on their body is not high on their daily agendas. Chimps need to pick out lice and other bugs. Dolphins do not. Perhaps that's why Delphi and Pan failed the mark (with no hands) test — perhaps neither one was sufficiently engaged by a foreign mark on his body. In which case, one could argue that the mark test was just unrealistically demanding. Or perhaps our experimental design was in some way inadequate.

In any case, I knew that before long I would have to try again.

6

THROUGH THE
LOOKING GLASS

THE DANCER IN ME was quite captivated as I watched Presley perform a bizarre sequence of horizontal swirls. He was lying below the water's surface on his left side, his body curled in a fetal position as he spun and looked, spun and looked, spun and looked. From modern dance lessons I had taken as a child, I knew that when a dancer executes a spin, he has to visually fix on a particular point after each rotation. It helps keep the dancer oriented and stable, and during practice sessions in front of a mirror it allows him to check out the aesthetics of the move. Presley appeared to be doing something very similar as he visually fixed on a particular point after each rotation. Spin and look, spin and look, spin and look. Round and round and round he went, this thirteen-year-old male bottlenose dolphin. In the five years I had known him, I'd never seen him (or any other dolphin, for that matter) do this, and he'd not been trained to do it either. It wasn't part of the natural behavioral repertoire of dolphins. Yet Presley was suddenly motivated to carry out the swirl. He was regarding himself in a three-by-five-foot horizontal mirror that I had placed in his pool. This dolphin spin-dancer glanced toward the mirror at the same instant in each rotation, looking at himself.*

---

* For video of Presley spinning and watching, see www.hmhbooks.com/dolphinmirror#spinning.

This was early in 1998 at the Wildlife Conservation Society's New York Aquarium in Brooklyn, and I had embarked on mirror-self-recognition investigations again, with two male dolphins, thirteen-year-old Presley and seventeen-year-old Tab, both of them captive-born. I had spent a lot of time thinking about the mirror we should use, trying to put myself in the mind of a dolphin. What would be the best possible demonstration that they really wanted to see themselves in the mirror? Then it came to me one day: *Make it smaller.* This way, if the dolphins truly wanted to see themselves, their actions would have to be quite deliberate. I thought we might see much more specific behavior as an indicator of their intentions. This approach would allow them to show me what they were capable of without my shepherding them toward a particular behavior.

In some ways it makes little difference what size a mirror is if all you are going to do is put an eye close to it, as I'd seen Presley and Tab do separately several times. But if Presley wanted to see himself fully in the three-by-five mirror while he swirled, he'd have to position himself some distance away from it, which is precisely what he did. He quite deliberately backed away from the mirror until he could see his entire body, and then he went into the horizontal swirl: spin and look, spin and look, spin and look. Presley apparently first tested out the physics of seeing the whole of his body in the small mirror, and then went into this entirely novel behavior, a move he invented.

⁓

Embarking on an ambitious research program with dolphins at the New York Aquarium hadn't been part of my plans when I left Marine World in Vallejo in 1990 for what I envisaged as a year's leave of absence for the birth of my daughter. My husband had a research position at Yale Medical School and we decided it would be best for me to join him there; we planned to stay there for a year and then return to the West Coast, where he would look for a professorship in the Bay Area while I would reunite with the dolphins and resume my position as director of the marine mammal lab. But it turned out to be a tumultuous year for the facility, and a fateful one for me. The administration changed. Amusement rides and other abominations were to be installed. Circe, to my piercing sorrow, was sold to an aquarium in Portugal, and Delphi and Pan were sold to an aquarium in Florida. The new director had decided he wanted to use the research pools as a breeding facility. Apparently, the new regime thought that Delphi and Pan didn't look enough like the prototypical Flipper and so were not suitable for dolphin shows. It's true that the boys were bigger than purebred Atlantic bottlenose dolphins, since they were a cross between the Atlantic and Pacific species. (Delphi's mother, Circe, and Pan's mother, Terry, were Atlantic dolphins, but the boys' father was the lovable, larger Pacific bottlenose Gordo.) But I thought they were adorable, and smart as hell.

I had left Brenda McCowan, who had been a graduate student in the lab and now boasted a PhD of her own from Harvard, in charge as acting director during my "temporary" absence, so she had some inkling of the seismic shifts that were taking place. I adore Brenda as a colleague and friend and hold her in high regard as a scientist. Our research team included Brenda, my lab manager Laura Edenborough, and many students from neighboring Bay Area universities and colleges, and collectively we created a family-like atmosphere at the lab. We cared deeply for one another, and for the dolphins and the research. I was very lucky. So it was understandable that Brenda and the other students tried to shield me from the unfolding bad news. When I eventually did find out what was under way, I was crushed. Losing the dolphins was like losing part of my family. In what was perhaps a pure fantasy, I tried, and failed, to raise funds to buy the dolphins myself. I intended to create a dolphin sanctuary of sorts. I know that "objective scientists" are supposed to be above such motives and emotions, but when you work closely with sentient, intelligent animals for years, strong emotional bonds are inevitable. When there is a connection of deep trust between an experimenter and animals they study and interact with on a daily basis, there is a greater likelihood of accessing the subtleties of the animal's mind. If that sounds unscientific to some ears, so be it. In my view, successful communication between human and nonhuman species is possible only through a genuine relationship.

You can imagine my state of mind: my Marine World family was broken up; my beloved dolphins were shipped off. It's as if a human mother had suffered a divorce and was permanently separated from her children. My husband, daughter, and I remained on the East Coast , and for the next half a dozen years I did what many young PhDs have to do: I became an academic vagabond, holding down short-term positions at various universities and colleges, often more than one at once. I endured crazy commutes while searching for the right opportunity to resume my research with dolphins. The institutions where I taught during these years were academically strong, including Yale, Columbia, and Rutgers. But I was driven to create my own dolphin research lab again. When an opportunity came to do that at the New York Aquarium, I took it, even though the facility was, shall we say, not ideal for the proper housing of the dolphins it already had. It was too small, and rather old. It was at Coney Island, right by the ocean, yet there was absolutely no sense of those surroundings inside. More than a little ironic for an institution housing large marine mammals. But with the encouragement of the second in command (and soon to be director), Paul Boyle, I put these reservations behind me. We shared the vision to transform the place into a much better facility for the animals, for research, and for public education about dolphins and their conservation and protection.

I first worked at the aquarium as a senior research scientist, testing the waters, so to speak, before becoming director of marine mammal research. There were few vacant offices, so I tried to be flexible. I installed myself in a small windowless area at the end of a hall with only enough space for a chair and a shelf as a desk. (A year later, when I became director, my research assistants and I moved into much larger offices and lab quarters on the second floor of the Osborn Laboratories of Marine Sciences at the aquarium.)

This was actually my second attempt to work at that aquarium. Two decades earlier, when I'd first made the decision to leave the theater and devote myself to dolphin research, I thought it might be wise to gain some experience with the animals before going to graduate school. I saw a help-wanted ad for a trainer to work with beluga whales at the New York Aquarium, and I applied for the job. As a marine mammal trainer I would have a lot of contact with the animals and would be able to learn a great deal about them. On top of that, the position required the trainer to be charismatic in front of an audience. Given my background in theater, I was completely comfortable with that. I was offered the job, and I accepted. Stuart and I made arrangements to move to Brooklyn from Philadelphia. We were poised to put down a rental deposit on an apartment, but the night before I was to begin the job, I got a call from the woman whom I was to replace. "I can't leave the job," she said tearfully. "It was a horrible mistake. I love the whales too much!" I surprised myself by being rather calm and accepting about what was really an awkward situation, and I said something like "Well, people do make mistakes, but it's okay." I took it as a sign that I should go to graduate school right then. I forgot all about that near job until the day I arrived in my office and the memories flooded back.

❧

There were two dolphin pools at the aquarium. One was a rectangular indoor pool, some sixty-two feet by forty-three feet and about ten feet deep; the other was an outdoor pool that the dolphins used during the summer months and that was in fact two connected pools, an oval one about forty-three by sixty-nine feet, again ten feet deep, and a smaller, round pool, twenty-seven feet in diameter. Three of the four walls in the indoor pool were glass, which allowed for public underwater viewing.

Under certain lighting conditions, from inside the pool these glass walls were somewhat reflective, like a hazy mirror; one corner in particular was quite reflective. The dolphins, Presley and Tab, had apparently noticed the windows' reflections, because soon after I started my research (another project on dolphin communication) I became aware that they were displaying several peculiar behaviors reminiscent of what I had seen Delphi and Pan do when they had a mirror in their pool at Marine World. These included close-eye viewing and adopting unusual orientations directed toward the walls, such as showing their bellies and displaying their penises. I frequently observed them doing this in the one tight corner where reflectivity was highest. I remember saying to Paul Boyle, who had become director of the aquarium, "It seems to me that these animals are recognizing themselves in the reflections." He asked how I could be sure. "Because," I said, "the things they are doing, putting their eyes close to the mirror, ventral presenting, these would be risky behaviors if Presley and Tab really thought the image was of another dolphin. They just wouldn't do that in social interaction with an unknown individual."

I thought to myself, *This is the right group to test for mirror self-recognition, because they are already experienced with mirrors.* I called Lori Marino and invited her to help me try another mirror self-recognition study. I explained what had prompted me, and she enthusiastically signed on.

Because of our frustrating experience with the mark (with no hands) test in Vallejo, Lori and I spent a good deal of time trying to figure out a different way to achieve the same end. We came up with another approach: We would train each dolphin to wear a small, gelatin suction cup, either black or white, on a part of the body where he could see it. (Edible gelatin ensured that if the dolphin swallowed it, it would just dissolve.) We would train each one to press a white lever when the suction cup he was wearing was white, and a black lever when the suction cup was black. When the dolphin became proficient at this, we'd put the suction cup on his forehead, where it would be invisible to him — unless he used a mirror. Would the dolphin use the mirror to check out the suction cup? And on seeing the by-then-familiar black or white cup, would he press the appropriate lever?

We thought it was a beautiful experiment and were quite proud of it. But then we realized it wasn't infallible as a demonstration of mirror self-recognition. It would be hard to explain away why the animal would go to the mirror after the suction cup was positioned on his head, but a skeptic could reasonably argue that the dolphin had learned: If I go to the mirror, I will see a suction cup, white or black, and then I will pull the appropriate lever. That line of reasoning did not necessarily imply self-recognition; it might mean that the individual was merely following a set of simple rules: see a white suction cup, pull the white lever; see a black one, pull the black lever. The dolphin in the mirror could easily be interpreted by the dolphin as a separate animal, and the lever-pulling would still be the same. So we had to give up on that. It would have been an easier experiment to run than the mark test, but, alas, it wouldn't do. So it was back to the mark test for us, with all its difficulties for no-handed subjects.

❧

We embarked on the mirror self-recognition tests in January 1998, setting the bar for success high. First, we predicted that if the dolphin knew it was his face in the mirror, he would not react as if he were seeing another dolphin — he would not show social behaviors. We also predicted that when marked after a feeding session, the dolphin would move to the mirror more quickly to check himself out than when he was not marked after a feeding session. But that alone was insufficient. The dolphin would also have to show himself highly motivated to inspect that part of his body that had been marked, clearly orienting that part of the body to the mirror as soon as he arrived.

We planned to videotape all the sessions and then score fifteen-minute segments comparing the behaviors the dolphins exhibited in the pre-mark and post-mark conditions. Four independent scorers in two independent pairs would rate the behaviors, second by second, as we watched the videotape segments. We coded behaviors in several different ways. *Mark-directed* meant that the dolphin positioned himself directly in front of the reflective surface and then oriented the body so that the mark could be seen in the reflection. *Self-directed* meant the dolphin was inspecting parts of his own body or watching his own behavior in the mirror (for example, close-eye viewing, looking inside his wide-open mouth, repetitive head circling, viewing his own genitals). *Social* meant the kind of behavior these dolphins would display in the presence of another familiar or unfamiliar dolphin, such as aggressive jaw-clapping and charging, or affiliative responses and vocalizations (whistling and squawking). Finally there was *exploratory,* meaning that the animal seemed to be inspecting the mirror itself, looking behind it or over it, or pushing on it.

As raters, we were unaware of whether each dolphin had been marked in any particular video segment being observed. We were "blind" to the condition we were rating. And we insisted that a particular instance of a behavior be included in further analysis *if and only if* there was 100 percent agreement between the two rating teams as to the behavior's classification. The coding was actually somewhat more complex than I've outlined, and it included not only the specific behavior at hand but also the time of onset of the behavior, time of termination, orientation of the body, and several other measures.

We also included several controls in the study. First, before we actually put a mark on either animal, we did sham marking. This involved taking the nontoxic dye out of the marker, replacing it with water, and then touching the dolphin on a part of his body he couldn't normally see *as if* we had truly marked him. We also planned to do the sham-marking procedure after the animals had experienced true marking. We hadn't done these controls in the earlier study at Marine World, nor had we been anywhere near as meticulous in analyzing post-marking behaviors.

The excruciating experimental details I've presented here are normally the stuff found in the methods section in academic papers, not in books for general readers. But our hypercompulsive and demanding experimental criteria were necessary because of what we were up against. If we announced that dolphins had the same cognitive capacities in the realm of awareness that humans and great apes *and no other minds in nature* had, then we had to back that up with iron-clad evidence. If we claimed that the primate-centered view of the origins of self-awareness in humans was incorrect, we had to have incontrovertible results. Extraordinary claims such as this one required extraordinary evidence. Gordon Gallup often says of his original work with the chimps that he felt he had demonstrated mirror self-recognition in the self-directed behaviors he observed, but he needed some more objective measure. That was the mark test. Lori and I felt the same about our study, but we had to work extra carefully with our non-handed dolphins.

As part of our extreme caution we decided to use only Presley as our principal experimental subject for the paper we planned to write. One of Tab's eyes was slightly cloudy. Lori and I talked about it and agreed that it would be prudent not to include him in the full mark test. We reasoned that Tab might fail the mark test because his vision might be impaired and not necessarily because he didn't recognize himself in the mirror. It was safer to leave him out, we decided. It later proved to be an unfortunate choice.

We carried out the first phase of the study, which involved getting base-line data on the dolphins' normal behavior in the rectangular indoor pool, before the mirror was installed. Then we exposed the dolphins to a very small, three-and-a-half-by-one-and-a-half-foot mirror (much smaller than the mirror described earlier) affixed vertically on the exterior wall of the pool, facing them. When the mirror was present, which was about half the time, the dolphins were definitely interested in it and in what it revealed. After a few initial bouts of touching the mirror, they began to position their bodies vertically, tightly aligned within the mirror's boundaries, as if fitting themselves into the reflective frame. Then each dolphin, alone at the mirror, began to do some very different, non-dolphinlike behaviors that were strikingly reminiscent of the antics of Groucho and Harpo Marx in the famous mirror scene in *Duck Soup* — except those comedians were crafting pretend reflections, with the straight man trying to catch the "reflection" in a mistake.

Like the Marx Brothers, the dolphins displayed a rich repertoire of some pretty bizarre orientations and highly repetitive movements — but only when they faced the mirror. For example, in calm silence, one of them would hang vertically within the narrow sliver of the mirror and repeatedly rock his body from side to side. Or he would move his head in wide arcs and circular movements close to the mirror, repeatedly and exaggeratedly nodding from left to right or up and down. Each would hang upside down at the mirror and peer into his wide-open mouth. The dolphins would do sequences of repeated flips. Like Groucho, and like our close ape cousins, the dolphins appeared to be testing the contingencies of their own behavior at the mirror. What happens when the dolphin circles his head? What does the other dolphin do? In mirror self-recognition studies, this is called the contingency-testing phase, a critical hurdle to pass on the way to self-recognition. It does not guarantee the next step of self-recognition, but it seems to be a necessary step for individuals who make that leap.

Presley and Tab did reach that milestone. They went on to

show clear signs of self-directed behavior. For example, they would look at the insides of their mouths, often holding their mouths open wide in prolonged gapes; they gazed at their eyes, holding their eyes right near the mirror surface; they blew varieties of bubbles at the mirror; they twisted and turned their bodies in bizarre postures (unlike their normal behavior in the absence of the mirror); and they brought their toys to the mirror and engaged in toy play there. The self-directed behavior at the mirror was a clear indicator that the dolphin understood that the external image in front of him, that dolphin in the mirror, was himself. In a young child, self-directed behavior is often reported as close inspection of eyes, mouth, genitals, or other body parts that go unseen without a mirror. Children might pick their noses, eat, or play in front of a mirror. If you mark each child in a group of nineteen-month-olds with a small spot of rouge on their foreheads or elsewhere on the face, about half of them will be able to look in the mirror and then point to the location of the spot on his or her body. Not all children of that age pass the mark test, although many show self-directed behavior, which is evidence in itself that they recognize the faces in the mirror are their own.

Both Presley and Tab were well used to being handled, tickled, and rubbed, and they often solicited it. They didn't seem to mind being marked when we did the initial sham marks. We didn't expect them to race to the mirror or any other reflective surface immediately afterward, and they didn't. We figured they would just think we were tickling or rubbing them. When the mirror was present, which was about half the time, they were definitely interested in it, observing themselves in various ways, but they made no attempt to examine the part of the body we had touched with the water-filled marker.

By October we felt we were ready to begin the true mark section of the study with Presley. The first time we marked him was quite revealing. As in the previous sessions, the dolphins were exposed to the mirror for thirty minutes, then called to their feeding areas to be fed and marked. We then gave them the release signal signifying the end of the feeding sessions, letting the dolphins know they could leave the area. This time, we carefully made a triangular mark on the side of Presley's head using a black, nontoxic marker. When we gave the release signal, he left station and swam around the pool playfully, as usual. When at last he swam by the mirror, however, he did a classic double take. He immediately turned around, went back to the mirror, and proceeded quite deliberately to orient himself in a way that would make the mark quite visible to him. We were elated.

We piled up statistically significant numbers of trials, and the results were outstanding, even more clear-cut than either Lori or I had dared hope. We were able to show that when marked, Presley swam right to the mirror and positioned his body immediately to expose the marked area to the mirror. We marked him on different parts of his body that he could not see without the mirror, and he always positioned himself accordingly. Orienting to the mark in his first moments at the mirror was strong evidence that he was using the mirror as a tool to view his mark. Presley spent substantially longer time in this behavior when he was marked than when he was not marked, more than twenty times longer. When he was sham-marked, he would swim over and inspect himself, but with nothing unusual to see, he would swim away much sooner. We had many qualitative observations too that made the same point: these animals knew who they were seeing in the mirror.

The dolphins' interest in the marks were never casual. They scrutinized their marks from many different angles, often for up to a minute. Presley demonstrated keen curiosity about the mark. We did most of these sessions in the two outdoor joined pools; the marking station was at the far end of the larger pool, and the mirror was around the corner from the entrance to the smaller pool. After being marked, Presley had to swim the length of the large pool, enter the smaller pool, and take a sharp left turn in order to use the mirror. He routinely did this very rapidly, sometimes in fewer than ten seconds, and never longer than a minute and a half.

In the summer, with the dolphins in their outdoor pools, we marked Presley under his chin on his neck area for the first time. We prepared to give the release signal, but before we could move an inch he darted away from his feeding station, made a fast beeline to the mirror in the adjoining pool, and assiduously examined the marked area. He arrived at the mirror so quickly that I could hardly keep him in my viewfinder as I videotaped his behavior. This time he oriented in a very different manner. He came close to the mirror and stretched his neck up repeatedly, lifting his head, exposing the marked area, and looking into the mirror. This self-examination went on for several minutes. Presley seemed very calm and very interested. Then he slowly backed away from the mirror and began to do the spin dance that I described earlier. If you want to see your entire body in a small mirror, you have to move away from it. He understood the rules. His behavior provided strong evidence that he was aware that what he saw before him was an external representation of himself.

We finally moved into what we called the late sham-marking phase, in which we again went through the motions of marking but with a water-filled marker. Presley had learned that after being touched by the marker, he was marked. As we expected, he raced to the mirror to inspect the spot. But he lost interest soon after seeing his usual, unmarked image. Presley spent about a fifth of the time in self-directed behavior inspecting himself when the mark was false and invisible.

On one occasion, though, Presley seemed determined to find a mark, even though there wasn't one. We sham-marked him on his right pectoral fin while he was in the rectangular pool and there was no mirror present. He immediately went to a corner of the pool that had the most reflectivity (due to a dark wall directly behind it) and spent more than half a minute in a sequence of a dozen dorsal-to-ventral flips in that corner area, each time bringing the sham-marked spot into close view. You could almost hear him thinking, *I'm sure they put a mark somewhere here.*

⁓⟨⟩⁓

I had seen several videos of chimpanzees undergoing the mirror self-recognition test and read many accounts of them, and it was obvious that Presley was at least as motivated as the chimps, if not more, to examine the marks in detail. The chimps usually touched the mark a few times, sometimes sniffing or tasting their fingers, but they fairly quickly lost interest. Not Presley, who was a mark self-examiner par excellence. Another difference between chimps and Presley was in the response to another individual's mark. While one chimp usually expressed interest in another individual's mark, neither Presley nor Tab paid any attention to the mark on his companion. Dolphins do not groom one another, as chimps do, so perhaps they have less interest in changes to another's appearance.

It's fair to say that anyone who observed Presley and Tab during this study would have been very quickly convinced that these animals were self-aware. I know that when I give public talks and show the videos of the dolphins after they've been marked, the audience sees what we saw. The tapes are compelling evidence that dolphins know there is a "me" there. But we needed quantitative evidence in support of that blindingly obvious qualitative conclusion. We probably could have published our results more than a year before we actually did. We knew we had it, but we also knew we had to convince everybody. So our two teams of raters spent more than a year grinding through the procedures I just described: detailed analysis of second-by-second behaviors, coding them by parallel independent teams. Not to put too fine a point on it, it was grueling, but there was no escaping it.

During the spring of 2000, Lori and I went through several drafts of the paper describing our results, and we sent copies to prominent colleagues in the field. We received very helpful comments and a lot of encouragement. We finally submitted what we felt was a finely honed manuscript to the journal *Nature* early in May. The British journal *Nature* is one of the most prestigious of scientific journals, and it prides itself on publishing breakthrough research across the spectrum of scientific disciplines. That was our rationale for sending it there: we felt that our work was a *really big breakthrough*. "This is a pivotal study that clearly demonstrates that the emergence of [self-awareness] is not a byproduct of factors specific to primates," we wrote in our cover letter, addressing the assumption that had been supported by a prodigious literature during the previous three decades. "These findings further advance our understanding of factors that may contribute to the emergence of advanced cognitive abilities in diverse species as well as providing us with a greater appreciation of the cognitive capacities of the [big-brained] dolphin."

We had high hopes for both a quick turnaround of the manuscript and its acceptance. We were right only about the speed. Within a couple of weeks I received a big brown envelope with *Nature*'s logo on it containing a firm letter of rejection and copies of reviews from four anonymous referees. We were completely deflated.

There's a joke among academic scientists about "the third reviewer," one that recently made the rounds on YouTube. Essentially, the idea is that you get two terrific reviews of your baby — the submitted manuscript — but then a third reviewer trashes your paper, and you have to rework it or do additional experiments. Most journals do indeed send manuscripts out to three reviewers. In our case, *Nature* sent it to four, and it was the fourth reviewer who killed us.

The first three reviewers were very positive, also offering helpful comments for improving the accessibility of the paper. One of the reviewers had scrutinized our statistics. This person judged them *unexceptional,* which is statistics-speak for "well within the norm." That referee also said that the effects we were testing with the statistics were so clear as to almost obviate the need for statistical analysis. Two of the referees commented on our small sample size of one — we had included only one individual, Presley. But researchers familiar with the field of animal cognition know that there is a long tradition of single-subject studies, such as those by Irene Pepperberg (with Alex, an African Grey parrot), Herbert Terrace, Allen and Beatrix Gardner, Duane Rumbaugh and Sue Savage Rumbaugh (all with chimpanzees), and many others. In scientific epistemology, the demonstration of a particular capacity in an individual of a species not previously known to possess that capacity is known as existence proof. It is the first step down the path of a particular discovery, and in our case we were confident that other bottlenose dolphins would eventually pass the mark test, just as we believed Presley had.

And then there was the fourth referee. This individual's tone was completely different, and the message was unequivocal: our paper should not appear in any scientific journal. The referee's criticism of our statistics was simply incorrect, but he or she had objections to other aspects of our study that struck us as emotional and knee-jerk reactions, clearly outside the norm of a professional scientific review. Lori and I were so convinced of the force of our scientific case that we took the unusual step of calling the editor at *Nature* to find out what we might do to get the decision reversed. He was very polite, very nice (he was British, after all), but equally firm. No, he could not publish the manuscript as it was. Perhaps if we had data from a second dolphin, he offered, he might be prevailed upon to reconsider.

We really didn't believe data from another dolphin was necessary, and we wrote a long letter explaining why. We also outlined the modifications we were willing to make in response to the first three referees' helpful comments and pointed out what we saw as the unambiguous factual errors of the fourth referee. To no avail. Our only way forward, it was made very clear, was to collect data on Tab and combine them with those from Presley. Why hadn't we done that in the first place? By this time, we were extremely stressed, having ridden the wave of exhilaration with our results, the devastation of initial rejection, optimism that our reasoned pleas would prevail, then rejection again.

In the midst of all this Gordon Gallup called me and said, "Congratulations on getting your paper into *Nature*, Diana." He had been one of the first three reviewers and he'd assumed that, armed with his very positive assessment, the paper would sail through. He was astonished when I said, "Well, actually, Gordon, it was rejected." He was very supportive and urged us to persevere. This was the scientist who'd developed the test we'd successfully put Presley through, and yet here Lori and I were, still knocking on *Nature*'s door.

We were able to collect the data we needed with Tab extremely quickly, for two reasons. First, we already had baseline data on him, in the form of a video record of his behaviors in the presence and absence of a mirror. And we had early sham-marking data on him too. Second, our experience with the mark test with Presley and our subsequent analysis allowed us to streamline our protocol. We knew what we were doing, and we were able to do it expeditiously. Less than a dozen actual mark tests and a handful of sham-mark tests gave us what we required. Tab's behavior was pretty much a carbon copy of Presley's, and we kicked ourselves more than a little for having thought that his possibly defective sight might interfere with his visual discrimination ability. Tab showed the same urgent motivation to get to the second, smaller, pool after being marked, and once there, he was just as diligent in examining the marked areas. There was no question, in our estimation, that Tab had the operational definition of mirror self-recognition.

We redrafted the paper and sent it back to *Nature* in the late summer with every expectation and hope that this time it would be accepted. After all, we had done what had been asked of us. Yet the paper was rejected *again*. The editor had sent it to the same four referees, and the fourth referee was even more strident in condemning the manuscript, essentially saying, I don't ever want to see this manuscript again. We were deflated once more.

There was nothing more to be done with *Nature,* and so we said, "Okay, let's try *Science.*" *Science,* based in Washington, D.C., is in many ways the American counterpart to *Nature,* and there is something of a professional rivalry between them. Like *Nature, Science* publishes groundbreaking papers across the spectrum of scientific disciplines, but the editors at *Science* are, if anything, even more conservative than those at *Nature.* Our manuscript bounced back to us from the editorial office of *Science* as if it had been attached by an elastic band, superfast. A curt rejection letter explained that the subject matter was deemed to be of insufficient general interest. We were advised to submit it to a more specialized journal.

This time we were not only disappointed but also astonished. Our experience over the years with dolphin research was that the public couldn't get enough information about dolphin behavior. Rather than go down the more-specialized-journal route, we chose *PNAS,* the *Proceedings of the National Academy of Sciences.* An inside joke in scientific circles is that *PNAS* also stands for "Post *Nature* and *Science.*" *Proceedings* is the house journal of the National Academy of Sciences, the most prestigious scientific institution in the United States. To have a manuscript accepted for publication in *PNAS,* authors must find a member of the academy who is willing to "communicate" the manuscript to the journal's editors. (Election to the academy is a much-coveted recognition that one has established great authority in one's particular field.) Like *Nature* and *Science, PNAS* publishes papers across a very broad spectrum of scientific disciplines, but these papers tend to be much more recondite than those of the other journals, not typically the topics of dinner-table conversation except among a small number of specialists around the world.

Lori and I discussed whom we might approach, and we quickly decided on Donald Griffin, the father of the study of animal cognition. I knew Don well, so I called him. He asked me to send him the manuscript and said he was excited about the results and would be happy to communicate it to *PNAS* and serve as the editor. The manuscript arrived at the *Proceedings'* offices on October 3, 2000, and was sent out for review. This time there was no negative third, or fourth, referee, and the paper was accepted; it was published on May 8, 2001. "These findings . . . offer the first convincing evidence that a nonprimate species, the bottlenose dolphin, is capable of [mirror self-recognition],"[1] we wrote. We also said that "research on self-recognition in other species will have profound implications for the idea that humans are the only species to conceive of their own identity."[2]

Lori and I were immediately swamped with requests for television, radio, and print-media interviews. Despite the incredulity of some reporters, who found it hard to believe that what we were describing could possibly be true, the media as a whole seemed entranced with the notion that bottlenose dolphins were "geniuses of the sea," as a reporter for the *New York Times* put it. The Wildlife Conservation Society (the parent of the New York Aquarium) has a press office that tracks media coverage of its scientists' work. I was astonished, and excited, when I got a memo from the office saying that we had had more than nine hundred media hits *in the first week.* It was by far the highest figure they had ever seen.

The story appeared on every major television channel in the country, and in every newspaper and newsmagazine. There was a spoof on *Saturday Night Live* in which the host asked what the difference was between a particular American pop star and dolphins. The answer: Dolphins are intelligent enough to recognize themselves in the mirror. On *The Daily Show,* Jon Stewart showed a photograph of me and said something like "Dolphins at the New York Aquarium are so smart that they recognize themselves in mirrors. Dolphin researchers aren't changing into their wetsuits in front of them anymore." Trivial Pursuit added a new question to the game: Which animal recently showed self-awareness?

My one regret about the *PNAS* paper is that we weren't able to include something Presley did twice, something that Lori and I thought was further corroborative evidence for MSR in the dolphins and something that people subsequently were curious about. "Did the dolphins ever try to rub off the mark after they'd seen it?" my scientific colleagues and other people often asked me. On two occasions after Presley was marked, he raced to the mirror, saw the mark, and then went to the side of the pool and rubbed the area of the mark against the pool wall. He then returned to the mirror, examined the location again, and seemed to be satisfied that he had achieved his goal. We had included this account in the first draft of the paper, but the paper was too long and we'd had to cut it.

This was especially unfortunate because a few months after the *PNAS* paper was published, a short item appeared in *Scientific American* entitled "The Flipper Effect." The Flipper effect, explained the small piece, was "the urge to believe that creatures as intelligent and engaging as dolphins must also be self-aware and empathetic." Primatologist Daniel Povinelli noted that dolphins didn't have hands with which to touch marks on their bodies but if they actively tried to rub off such marks, he would consider that evidence of self-awareness.[3] Until such time as this was observed, however, he would remain unconvinced.

Nonetheless, we felt we had achieved our goal of meeting the rigorous standard that Gordon Gallup had set three decades earlier with his chimp work. Dolphins had shown themselves to be capable of self-recognition and this level of self-awareness. They have consciousness, just as chimps and people do. When I think about how difficult it was to establish this, what comes to mind is Bob Thaves's quote about Ginger Rogers: "Sure [Fred Astaire] was great, but don't forget, Ginger Rogers did everything he did backwards . . . and in high heels!" Well, Presley and Tab had done everything that Gordon's chimps had done.

And with no hands!

7

COGNITIVE COUSINS

DAAN, A MATURE MALE bottlenose dolphin, shared a very large pool for many years with several other dolphins and a mature male Cape fur seal in an oceanarium in Humewood, South Africa. Daan was an inquisitive individual, as dolphins usually are, and he became particularly interested in the divers that from time to time entered the pool to remove algae from the inside of the glass viewing ports. He spent long periods patiently watching the divers as they scraped, scraped, scraped at the film of algae on the glass. Very soon, Daan took it into his mind to help out. He found a seagull feather, held it in his mouth, and proceeded to clean the window just as he had seen the divers do.

When a diver did the cleaning, he steadied himself by holding on to the frame of the viewing port with one hand, and he used the other to clean the glass with broad strokes of a brush. Daan adopted the whole posture. He kept constant contact with the frame using one flipper, and he used the feather to clean the glass with the same broad strokes he'd seen the divers use. And he hadn't just watched the divers; he'd also listened to them. As he cleaned the windows he emitted "sounds almost identical to that of the diver's air-demand valve" and he released "a stream of bubbles from the blowhole in a manner similar to that of exhaust escaping from the diving apparatus."[1] Daan was apparently so enamored with his "job" that for almost two months he refused to allow any diver near the window, and at night he rested above it rather than in the center of the pool, which had been his previous nocturnal spot.

Mimicry of this sophistication requires a good deal of intelligence. And dolphins don't have to be mature to display such inventiveness. Dolly was a six-month-old bottlenose dolphin in Daan's pool. She loved the attention of the observers who were scoring dolphin behavior in an underwater viewing chamber. Dolly developed the habit of offering the observers objects, such as feathers, stones, seaweed, and fish skins, pressing them against the glass as tokens of the relationship she seemed to want to establish. When the observers ignored Dolly, she would rapidly bring three or four objects in quick succession. Dolly's endearing behavior was hard to ignore.

One day, one of the observers was smoking a cigarette and happened to let out a large plume of smoke. Dolly apparently noticed it, and without hesitation she swam to her mother, briefly suckled, returned to the window where the observer was watching, and let out a mouthful of milk that "engulfed her head, giving much the same effect as had the cigarette smoke."[2] The observer reported that he was "absolutely astonished," and he must have displayed his feelings, because on several subsequent occasions, Dolly used this little improvised display to attract attention.

Had Dolly occasionally "spilled" milk while suckling and remembered what it looked like? Perhaps, though I'm sure it doesn't happen often, to judge from my observations of mothers and calves. Had Dolly made the connection between the visual appearance of the cloud of smoke and the cloud of spilled milk and then used that recognition in a completely novel way? Imitation is one of the highest forms of social learning, and it requires selective attention and considerable intelligence. Dolphins are extremely good at it, even better than chimpanzees. This display of intelligence prompted primatologist Andrew Whiten to observe that dolphins "ape better than apes."[3]

These two stories are wonderful examples of dolphins actively observing us while we're observing them. I know from experience that you don't have to be around dolphins for very long before you become keenly aware that you are an object of dolphin scrutiny and that there's a palpable intelligence behind it. These anecdotes were both reported by a trained observer, the aquarium curator, who was familiar and knowledgeable about the behavioral history of the dolphins. Anecdotal reports by trained observers when compiled and viewed as a whole can give us important glimpses and insights into the minds of other species and can inform new directions of scientific investigation into those species' cognitive realms. The anecdotes don't generally get published in scientific journals, but they are telling nonetheless.

The questions generally asked are: Just how intelligent are dolphins? Are they almost as smart as humans? Are they smarter than chimpanzees? Yet the question of who is more or less intelligent is perhaps the wrong one to ask. As a young graduate student I found Ross Ashby's definition of intelligence very helpful. Ashby, a cybernetician, studied complex systems and defined *intelligence* as the "power of appropriate selection," be it in man, animal, or machine. Ashby's definition levels the playing field and allows intelligence to be considered relative to the sensory abilities and social and environmental factors at work. He proposed that

> "problem solving" is largely, perhaps entirely, a matter of appropriate selection. Take, for instance, any popular book of problems and puzzles. Almost every one can be reduced to the form: out of a certain set, indicate one element . . . It is also clear that many of the tests used for measuring "intelligence" are scored essentially according to the candidate's power of appropriate selection . . . Thus it is not impossible that what is commonly referred to as "intellectual power" may be equivalent to "the power of appropriate selection." Indeed, if a talking Black Box were to show high power of appropriate selection in such matters — so that, when given difficult problems, it persistently gave correct answers — we could hardly deny that it was showing the "behavioral" equivalent of "high intelligence."[4]

Dolphins are very clever at locating and capturing schools of fish. Humans do the same thing by cleverly designing and using boats and nets. Every intelligent species (and others that are less well endowed mentally) will, because of its unique evolutionary history, possess special abilities unmatched by other similarly intelligent species. As the University of Wisconsin psychologist Charles Snowdon put it, "Specializations do not make one species 'smarter' than another, but they do make for uniquely sculpted minds."[5] Finding the answer to the question, Who is smarter than whom? is therefore not always helpful, or even possible. When we ask that kind of question, we inevitably think about the issue in very human-centered terms, based on our notion of intelligence.

Dolphins are a large-brained, highly gregarious species whose body form and sensory systems have been exquisitely shaped for life in the seas by millions of years of evolutionary selection. Working with them has presented me with a gnawing concern: Would we recognize their form of intelligence, the meaning of their messages, even if it were right there in front of us? These nonprimates, so alien in body form, from so foreign an environment — how do we search for meaning in their behavior and communication? What may be meaningful to dolphins, so important to them in complex sea-based social networks, may be so alien to us that we can't even conceive of it.

Ludwig Wittgenstein famously said, "If a lion could talk, we wouldn't be able to understand it." Wittgenstein's point was that even if the lion could speak perfectly good English, its experience of the world is so different, so alien, compared to ours that what the lion would have to say would make little sense to us.

To some degree, there is the same conundrum when we think about another species' intelligence. We have to ask ourselves whether we would even recognize an intelligence in another species that is radically different from ours. Douglas Adams nicely captured the essence of the problem in *The Hitchhiker's Guide to the Galaxy*:

"It is an important and popular fact that things are not always what they seem. For instance, on the planet Earth, man had always assumed that he was more intelligent than dolphins because he had achieved so much — the wheel, New York, wars and so on — whilst all the dolphins had ever done was muck about in the water having a good time. But conversely, the dolphins had always believed that they were far more intelligent than man — for precisely the same reasons."[6]

It is clear that dolphins are among the smartest creatures on Earth. So what do we know about dolphin minds compared to other minds? What parallels do we see with other intelligent species? What insights can we glean from these observations about the nature of intelligence and other high-level cognitive functions, such as self-awareness? I *know* dolphins are intelligent based on what I've seen and experienced with them over more than three decades. The question is: How do we characterize their *kind* of intelligence? In other words, what is it like in the mind of a dolphin?

Let's start with something a bit more concrete. Let's talk about brains. Humans, in contemplating our place in nature, have long been obsessed with our species' big brain. We named ourselves *Homo sapiens,* "wise man" or "knowing man," in recognition of the powers with which it endows us. The human brain is held to be superior relative to other species' due to its larger-than-expected size and its enlarged, well-developed cerebral cortex. Brain size has been linked to intelligence because in mammals and birds, those species with the largest brains show the greatest range and versatility in behavior.[7] Dolphins and other toothed whales also boast big brains, significantly larger than our own. The bottlenose dolphin — Pan and Delphi's species — has a brain that weighs about seventeen hundred grams; the human brain weighs about thirteen hundred grams, so the dolphin brain is about 30 percent larger. Much of the dolphin's brain weight is made up of cerebral cortex — considered to be the seat of higher cognitive functions.

If you are developing something of an inferiority complex, you might comfort yourself with a measurement called the encephalization quotient, or EQ.[8] This measure takes into account the size of one's body relative to the size of one's brain. EQ represents the *ratio* of brain weight to the average body weight for a given species. Among mammals, for example, *Homo sapiens* has an EQ of 7.0, which means that the human brain is seven times larger than expected for the human body size. Our closest biological relatives, the great apes, possess brains that are also larger than expected, a bit more than twice as large, but they are relatively smaller brains by this measure: the EQ for chimpanzees is 2.3; for orangutans, 1.8; for gorillas, 1.6. Elephants and whales also have larger brains than expected, and they show a ratio similar to that of the great apes: more than twice the expected size. Most other animals have smaller brain-to-body ratios than humans and great apes. With one exception — the dolphin. Dolphin brains are approximately 4.2 times larger than expected for their body size — the highest ratio of any species other than humans, making them easily the second most cerebrally endowed species on the planet, way ahead of the great apes.

Another measure of intelligence and brain complexity is based on the amount of cortical folding. If you have an eight-by-eleven-inch sheet of paper and want to fit it into your closed fist, you would scrunch it up, causing the paper to crease and fold in on itself. This scrunching (to use a technical term) allows increased surface area within the same small volume. So a greater degree of cortical folding, or convolutions, represents more surface area, and with increased surface area, one can get a much larger number of brain cells into the brain without changing its volume. This is important because the total number of brain cells, or neurons, may also be a factor in intelligence. The part of the brain that is convoluted is called the cerebral cortex, that part of the mammalian brain associated with higher cognitive processes and thinking. Our brains are more convoluted than those of our primate relatives. Only one brain is known to be more heavily convoluted than ours. Again, it's the dolphin brain.

Recent comparisons of different measures such as brain size and relative brain size have led some scientists to conclude that the average number of neurons in the brain may correlate better with intelligence when making comparisons across species.[9] Recently Suzana Herculano-Houzel and her colleagues used a novel technique they developed to count neurons in the human brain and in those of other species. They estimated that the human brain has approximately eighty to ninety billion neurons; the great ape brain has approximately twenty-four to thirty-two billion neurons; the elephant has around twenty-three billion neurons; and the false killer whale has about thirty-two billion neurons. No data for dolphins are in as of yet, and it will be interesting to see how they compare using this measure.

But an additional factor to consider is the size of the neurons themselves in different brains. Our neurons and those of other primates are smaller than those found in the brains of dolphins, whales, elephants, and rodents, and therefore more brain cells can fit into a given space in the human and primate brains. It has been suggested that there are two different scaling schemas: a primate scheme, with smaller neurons but more of them, and a rodent scheme, also found in dolphins, whales, and elephants, in which there are larger neurons but fewer of them. These may represent two different architectural strategies for building a brain.

Even neural numbers and brain size together may not provide the whole picture. For example, insects with very small brains learn and perform complex behaviors with a very limited number of neurons. This suggests that computational power may depend not only on the size but also on the organization of the brain — the way in which the neurons are connected. Even the specific types of cells found in brains may provide clues about intelligence. Von Economo neurons (VENs) are specialized, elongated spindle-shaped cells that are found in specific parts of the human brain and are thought by some to be linked to social cognition, empathy, emotion, theory of mind, and "gut" feelings.[10] Patrick Hof at Mt. Sinai Medical School and Stuart Allman at Caltech and their colleagues have discovered VENs in specific regions in the brains of humans, great apes, elephants, and cetaceans — those species showing complex social and cognitive abilities, including the capacity for self-awareness (as indicated by mirror self-recognition) and empathy.

Taken as a whole, we can no longer rest on the assumption that the human brain is unique and superior to those of other animals based on its absolute or relative size, the size of the cerebral cortex, or its degree of convolutedness. This is certainly not to suggest that the human brain is not exquisite in its capabilities. Instead, it helps us to realize and appreciate that other brains, other minds, have evolved in parallel to meet complex social and environmental challenges. Other exquisite minds may exist as well. Social minds, self-aware minds, sentient minds in the seas.

∝∞∽

Metabolically, the brain is a very expensive organ to maintain. In an adult human, the three pounds of gray and other neural matter in the head represents less than 2 percent of total body weight, yet it consumes about 18 percent of the body's energy budget. And building a large brain in the womb is a burden for the gestating mother. Large-brained species can sustain their elevated cognitive status only if they have diets rich in energy. Nature is a very conservative mistress, and she maintains energy-expensive organs and behaviors only if there is some selective advantage for doing so. If a species is burdened with the exceptional energy demands of a large brain, then being brainy must therefore have exceptional benefits. The question then is, Why do dolphins need to be as smart as they are? That is, what are the benefits of having a powerful mind in the waters?

Let us step back a minute and think about our own cognitive

endowment. We live in a world where our fellow humans write computer code and symphonies, smash atoms and think about weird quantum worlds, create wondrous pieces of theater, art, and architecture, and imagine florid fantasies such as *The Hitchhiker's Guide to the Galaxy*. And yet the brains that accomplish these remarkable feats of creative intelligence are the same brains that evolved to meet the cognitive demands of a nomadic, small-band, hunting-and-gathering lifestyle in the Paleolithic period, more than a hundred thousand years ago. The men and women back then had the same type of brain as the average university professor today, and yet they spent their lives stalking and killing antelope and other such game, collecting berries, and digging up tubers.

In his book *Mind from Matter,* Max Delbruck, a Nobel laureate in physics and an early molecular geneticist, discussed our large human brains and wondered why so "much more was delivered than was ordered."

We can ask the same kind of thing about dolphins. As we know now, they have the ability to learn arbitrary symbols and use them appropriately in relation to specific objects. Being able to respond appropriately to arbitrary hand gestures and being sensitive to "word" order, as dolphins are in Lou Herman's facility in Hawaii, is of the same intellectual level (see next chapter). The dolphin literature is full of instances where these animals perform tasks that require not only motivation and attention but also complex problem solving and creative intelligence.

These experimental activities have something unusual in common: they are carried out under circumstances that are, cognitively speaking, completely foreign to dolphins' daily lives. Dolphins show us that they can operate with great skill when faced with intellectual challenges that are alien to a dolphin's life in the ocean. Yet they handle such arbitrary and abstract concepts and tasks with ease. Pan and Delphi and others in my research lab, and Akeakamai and Phoenix and other dolphins in Lou Herman's lab, performed tasks that appeared to be different than anything their natural environment might have prepared them for. How come they could do these things?

How come humans and dolphins have brain capacities that appear to exceed the daily exigencies under which they evolved? And chimpanzees, and perhaps elephants too, for that matter? If we put the question in the terms of evolutionary biology, it becomes: What selective advantage does a high degree of intelligence confer on humans, dolphins, chimpanzees, and elephants in their state of nature?

We have long been obsessed with the forces that drove the big human brain along its evolutionary path. Language has often been mentioned as contributing to the large cerebral cortex — but there is little evidence that language centers in the brain account for an unusually large fraction of brain mass. In all probability, brain expansion in the early human lineage began long before spoken, referential language came into wide use. Human brain expansion was under way by about two and a half million years ago, around the same time that stone tools started to appear in the archaeological record. (Earlier than that, the brains of our human ancestors were pretty much apelike in size.) In the 1950s, Man the Toolmaker[11] became a popular model to explain the increase in brain size in early human evolution. A decade later, this model was replaced by Man the Hunter,[12] as anthropologists became fascinated with technologically primitive societies and what they could tell us about how our ancestors might have lived. (Both of these models are very much male oriented, of course: Man the [Something Macho]. Some balance was restored when an alternative, Woman the Gatherer, was offered in the mid-1970s.*,[13])

Whatever their separate merits, both of these models focused on the world of practical affairs, of subsistence, as the evolution-

---

* This latter probably reflected much more what our ancestors' hunter-and-gatherer life was like, as in the !Kung San, of Botswana, until not so long ago. Women provided most of the dietary sustenance, the nuts and tubers, whereas the men only occasionally brought home some meat.

ary engine of brain expansion. Yet at about the same time that the Woman the Gatherer model was attracting a lot of interest, an important shift of emphasis began to take place, propelled by the paradox I outlined above: in laboratory situations, chimpanzees appear to be too smart for their own good. "Some primate species (and mankind in particular) are much cleverer than they need be," Nicholas Humphrey, a psychologist at Cambridge University, wrote in a now classic paper called "The Social Function of Intellect,"[14] published in 1976.* Surprising as it may seem, until this point biologists had not given much thought to how intellect might contribute to biological fitness; that is, success in producing offspring. Humphrey was promulgating a shift in perspective from the world of practical affairs to the world of social affairs and the intellectual demands of managing relationships in complex societies.

When biologists did address the function of intellect, or creative intelligence, it was mainly in the context of making and doing things. For humans, this included inventing new tools and finding new ways to use existing tools in order to make better use of environmental resources. Although Jane Goodall discovered in 1960 that chimpanzees also make and use tools, for them it is an extremely limited activity. Great apes are hardly simian handymen! Nor are they especially energetic or inventive when it comes to gathering their daily vittles, as Humphrey observed:

"The great apes, demonstrably the most intellectually gifted of all animals,† seem on the whole to lead comparatively undemanding lives, less demanding not only of those of lower primates but also of many non-primate species. During the two months I spent watching gorillas in the Virunga Mountains I could not help being struck by the fact that of all the animals of the forest the gorillas

---

* Princeton biologist Alison Jolly had published similar ideas a decade earlier, but they didn't have the impact on thinking among psychologists and primatologists that Humphrey's paper did. See Alison Jolly, "Lemur Social Behavior and Primate Intelligence," *Science* 153 (1966): 501–6.

† This paper was written before the extent of dolphins' cognitive abilities was known.

seemed to lead much the simplest existence — food abundant and easy to harvest (provided they *knew* where to find it,) few if any predators (provided they *knew* how to avoid them) . . . little to do in fact (and little done) but eat, sleep, and play. And the same is arguably true of natural man."[15]

While the world of subsistence for the great apes is relatively undemanding and predictable, the world of interpersonal interaction in complex social systems is anything but. And this was the nub of Humphrey's insight: "I propose that the chief role of creative intelligence is to hold society together,"[16] he wrote. The overall structure of chimpanzee society is known as fission-fusion, which means that from time to time small groups come together to form a larger group that eventually splits apart into the original subgroups, with some change in group membership occasionally taking place. Mothers and offspring (females and younger males) are at the center of chimpanzee society; older juveniles form their own groups; and older males are sometimes solitary and sometimes band together for hunts or even attacks on neighboring groups. When young males reach maturity, they leave to find another group, where they spend a lot of social skills being accepted as members.

The organizing principle of chimpanzee society, indeed of all animal societies, is reproductive success. Males attempt to sire as many offspring as they can, while the females' goal is to be courted by the most genetically desirable males. In most animal societies, the outcome of a challenge by one male to another (for access to females) is rather predictable. The winner is the bigger male, or the one with longer canines, or the bigger antlers (or whatever weapon of male-to-male combat is appropriate). Not so for chimps, baboons, and other large monkeys. Although physical prowess is helpful in these higher primates, an individual male's ability to form friendships or alliances with other individuals, both male and female, is key to reproductive success. A weakling male can sometimes mate with a desired female, provided he times his amorous advances well and moves in when his friends are at hand to help him fend off a challenge by a more dominant male or when the other male's allies are not around to intervene.

Being socially adept in a complex social group therefore requires remembering who is related to whom, which individuals have recently formed an alliance, whom you have helped recently and therefore might expect help from when the need arises, and so on. The intellectual challenge is made greater by the constant shifting of alliances, as other individuals in the group change their allegiances, in hopes of greater advantage. A shift in allegiance by a single individual might subtly change the balance of power, causing further changes to cascade throughout the group.

Every individual, in order to maximize his or her reproductive success, is constantly calculating and recalculating the balance of power in the group among older and younger uncles and aunts, nieces and nephews, and unrelated individuals too. It's a very complicated game with a constantly changing set of rules.

These multigenerational groups provide protection and an environment in which the young can learn subsistence and parenting skills. Yet the social challenges were novel in the world of nature when they arose in higher primates. "It asks for a level of intelligence which is, I submit, unparalleled in any other sphere of living,"[17] wrote Humphrey. With the evolution of higher-primate societies, individuals had to become what Humphrey called "nature's psychologists." And once social skills become an effective element in the equation of reproductive success, a feedback loop arose:

"If intellectual prowess is correlated with social success, and if social success means high biological fitness, then any heritable trait which increases the ability of an individual to outwit his fellows will soon spread through the gene pool. And in these circumstances there can be no going back: an evolutionary 'ratchet' has been set up, acting like a self-winding watch to increase the general intellectual standing of the species."[18]

Nature's psychologists need to be more intellectually gifted than creatures whose social environment is less complex. Humphrey called this the social intelligence hypothesis[*] and used it to explain the evolution of big brains relative to body size. This line of reasoning quickly became popular (and still is), and was co-opted in a way by Richard Byrne and Andrew Whiten, primatologists at the University of St. Andrews; they preferred the phrase the *Machiavellian intelligence hypothesis*. Niccolò Machiavelli (1469–1527) was an Italian philosopher whose most famous work, *The Prince*, instructs the reader in the ways of social and political success through clever manipulation of relationships and alliances. Many people, myself included, believe the phrase Machiavellian intelligence sounds unnecessarily negative. I prefer the term *social intelligence* or *social cognition* — the skills are related to attraction and cooperation, not just competition.

Humans, like chimps, live in a fission-fusion society. So do bonobos (pygmy chimpanzees), several other primate species, and many nonprimates, such as lions, deer, and even some fish. The dolphins' social organization is also fission-fusion.

Dolphins form close and long-lasting bonds with one another that can last lifetimes, and they often interact collaboratively and cooperatively in alliances in their myriad of foraging strategies, in mating, and in the rearing of young. Alliances can last from minutes to hours or can be long-term relationships. For example, three male dolphins may spend the majority of their time together swimming and finding food in what is called a first-order alliance. At times they may rejoin their larger social group or mix and mingle with other individuals in other alliances and form new, more temporary alliances. Related or unrelated females with young calves form another type of alliance and spend time together in subgroups collectively caring for their offspring. Dolphin mothers, like human mothers, seem to have to learn about parenting skills, and depend-

---

[*] Alison Jolly had termed it the *social use of intelligence*.

ing on their knowledge and disposition, they show varying degrees of vigilance toward their youngsters. I have observed a wide range of mothering skills and styles over the years. Females will often allomother — baby-sit and care for another's calf, allowing the real mom to forage or rest a bit. Dolphins can spontaneously begin to lactate and, like wet nurses, provide needed nourishment for an orphaned calf.

Richard Conner and his colleagues have reported that bottlenose dolphins living in Shark Bay, Australia, form multiple-level male alliances within a social network; they suggest that this level of alliance complexity has been found previously only in human societies.[19] If you're a male dolphin living in Shark Bay, you may cooperatively herd females with the one or two members of your first-order alliance — your closest buddies. But you may also join up with another alliance to form a larger second-order alliance, or even meet up with other alliances to form a multiple alliance of males that cooperatively work together to compete with other multiple alliances of males in herding and guarding females with whom they hope to mate. Alliance formation and cooperative behavior is considered a hallmark of social complexity and requires sophisticated social cognition.*

Where does self-awareness, or consciousness, fit into this hypothesis? Are nature's psychologists really just automatons, animals with "clever brains, but blank minds"?[20] Or are they creatures that are aware of their actions and feelings, conscious of themselves in the world? Does self-awareness make a more effective social individual? Many of these ideas are showing up in studies of consciousness, an area that until recently was avoided by neuroscientists be-

---

* If you think this sounds complicated, it is. Think of your own social network. You have close friends; acquaintances whom you mingle with on occasion; friends of friends whom you may socialize with; parents of your children's schoolmates; professional affiliations that accomplish specific goals; reunions of organizations; and large familial gatherings. It takes a high level of cognition to keep track of and remember all one's social and familial affiliates.

cause it seemed too difficult to quantify and experiment with. This is changing.

⚬◯⚬

First, a word about consciousness, a topic that has long enthralled and bewildered philosophers and biologists alike. To certain philosophers, consciousness is a private phenomenon, something we humans enjoy as individuals, giving one a sense of self and sometimes even an experience of transcendence. But according to this line of thought, consciousness doesn't materially contribute to physical well-being. And because it is private, consciousness (in humans and, if it exists, in animals) is therefore necessarily inaccessible to the forces of natural selection and to scientific investigation.

To most biologists, however, consciousness is presumed to be a beneficial trait, the product of evolution by natural selection. If so, consciousness cannot be entirely a private thing, because it produces behaviors that are visible to natural selection. Individuals with well-developed consciousnesses should have an advantage over those who are less gifted. If this is true, it means that consciousness will become more sharply honed in the population over time, an evolutionary ratchet of a sort. Consciousness should, in principle, be accessible to scientific inquiry. But as you saw in the mirror self-recognition work with Delphi and Pan, and with Presley and Tab, self-awareness resists easy detection, at least in an objective, scientific manner. And is self-awareness the same thing as consciousness? I know I am conscious, because I experience "me." And I assume you are conscious too, because you are a fellow human being. But *assumptions* about nonhuman animals are much more tenuous, even though one might viscerally *feel* the presence of consciousness in a nonhuman, sentient animal.

Nonetheless, let's consider what the utility of consciousness could be, why it would have evolved through natural selection. Or, to put it another way, How does it benefit a human or a dolphin to look into a mirror and recognize the image as one's self?

Although consciousness is what we experience as self, most of what goes on in our minds is veiled from our conscious experience. Most of the brain's activity is concerned with maintaining the myriad physiological and motor systems, keeping the biological machine in good working order. We have no need to be conscious of the constant monitoring of activities and the requisite activation signals. Even an action that requires intense concentration to learn, such as riding a bike (which demands, among other things, constantly monitoring balance) or serving the ball in tennis (which requires careful monitoring of how the ball is thrown into the air and how to swing the racket), become automatic once one becomes proficient. What had been a conscious effort no longer is. And many a person has had the experience of being at the wheel of a car and suddenly realizing that he has not been actively aware of driving the car for the past couple of miles. The conscious mind was "elsewhere" while the unconscious mind was dealing with the practicalities of driving safely along the road.

All of our active behaviors — getting out of bed in the morning, cooking breakfast, going to work — can be accompanied by conscious feelings and explanations concerning what's happening: wishing to stay in a warm bed for just a few more minutes; feeling hungry; feeling the need to make a living. In other words, the conscious mind has the capacity to monitor one's state at any particular moment of the day and to report it to the self. But, as with the "driving blind" experience, it is possible to imagine doing all the above activities on autopilot, without emotions or explanations coming to mind.

Even the smartest of species with the cleverest of brains often operate on cognitive autopilot, deftly navigating the daily practicalities of subsistence and social interaction while beyond the realm of awareness. How, then, is consciousness an advantage? It's the first step in social reasoning. An individual who is aware of his own actions and emotions under particular circumstances is, in principle, able to predict the actions and emotions of another individual under those same circumstances. In the game of social interactions, as among the complex social systems of chimps, self-awareness confers a distinct advantage. An individual endowed with consciousness would be able to navigate the ever-changing web of relationships among the alliances within its community far more effectively than a behaviorist. Once the faintest spark of consciousness arose in the minds of individuals in a species, the forces of natural selection would inexorably fan it, generation after generation, so that eventually it would glow brighter and brighter. Ultimately that species would possess an aspect of what has been called the theory of mind — the ability to predict what is in the minds of others based on one's own experience.*

In the literature of evolutionary biology, the discussions around the existence and utility of consciousness among nonhumans focused exclusively on higher primates, notably the great apes and some of the larger Old-World monkeys, such as various species of baboon. This primate-centered focus was presented by primatologists David Premack and Guy Woodruff in a landmark paper titled "Do Chimpanzees Have a Theory of Mind?"[21] published in 1978. In the introduction to the paper, Premack and Woodruff said, "An individual has a theory of mind if he imputes mental states to himself and others." Which is an appropriately academic way of saying "I have an idea of what's on your mind, because I know what's on mine."† Premack and Woodruff answered their own question affirmatively: yes, chimpanzees do have a basic theory of mind.

By the 1980s, most biologists agreed that chimpanzees have the

---

* Interestingly, the English philosopher Thomas Hobbes (1588–1679) introduced this argument three centuries ago in his book *Leviathan*: "Whosoever looketh into himself and considereth what he doth, when he does think, opine, reason, hope, fear, etc., and upon what grounds; he shall thereby read and know what are the thoughts and passions of all other men upon the like occasions."

† Theory of mind is a great deal more complex than this, of course, and includes the ability to know that another can have a belief about the world that is false.

kinds of minds they have because of the highly complex social environment in which their ancestors evolved. Individuals produced more offspring if they were socially adept; social intelligence — the ability to correctly assume what was going on in the minds of others — was a powerful tool. *Homo sapiens* occupied the peak of this mountain, but the great apes were allowed to camp in the foothills; they were not as cognitively endowed as us, of course, but they were above the common ground of the rest of the animal world.

Do dolphins have creative intelligence, as chimpanzees do? Absolutely. Do dolphins have a theory of mind to the extent that chimpanzees have? I have no doubt that they do. Dolphins are, in the words of Louis Herman, "cognitive cousins" of chimpanzees.[22]

As it turns out, dolphins have other cognitive cousins, in addition to humans and chimpanzees. Gordon Gallup first suggested that dolphins might pass the mirror test because, like chimpanzees, dolphins have large, complex brains, lead complex social lives, and show evidence of empathy toward others. Gallup reasoned that these factors correlated with the ability for mirror self-recognition in chimpanzees, so he was not surprised when Presley and Tab passed the mirror test.

Elephants also have large brains and show empathy, so a few years after we completed the test with dolphins, I teamed up with Emory University primatologist Frans de Waal and his graduate student Joshua Plotnik and tried the mirror test with three adult female elephants at the Bronx Zoo. Josh and I spent hot summer days atop the roof of the elephant house filming the elephants' behavior when they were in front of a jumbo and very strong eight-by-eight-foot mirror. This time, although the animals had no hands, they did have trunks. Remarkably, the elephants and dolphins showed strikingly similar behaviors and went through the same three stages as the chimps had upon their first encounter with mirrors. Happy, one of the three elephants, passed the mark test by using her trunk to pat the mark we had placed on her head.[23]

We submitted the manuscript reporting the study to both *Na-*

ture and *Science,* but neither journal even sent it out for review. Again, it was deemed of insufficient interest for the general public. The paper was published shortly thereafter in *PNAS,* and as with the dolphin paper, the story was widely covered in the media. My favorite media coverage was by Associated Press writer Andrew Bridges, who opened his article with "If you're Happy and you know it, pat your head." I now tell audiences that I work with "big gray animals with big gray brains," and they are all smart as hell! (See figure.)

Some have argued that dolphins' brains need to be as remarkably big as they are in order to facilitate the dolphins' superb echolocation abilities, which exceed even the U.S. Navy's most advanced equipment. Perhaps this impressive biological skill requires an unusual amount of brainpower. Yet there's no evidence to support this idea. Bats, with much smaller brains, have a comparably sophisticated natural sonar system. A controversial suggestion was made more recently: perhaps much of the dolphin's brain is made up of thermogenic glial cells, which would have helped dolphins survive the decrease in ocean temperatures that occurred about forty million years ago. In this view, the dolphin's large brain is an accidental

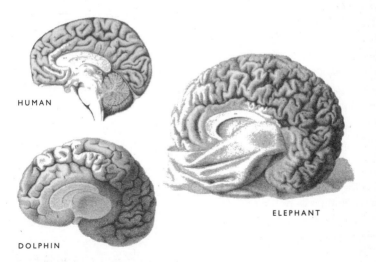

HUMAN

DOLPHIN

ELEPHANT

benefit of its having needed a kind of internal electric blanket. But the most widespread view is that the brains of dolphins and other cetaceans evolved to support complex cognitive abilities; this is the social competition thesis. It's certainly my view.

Just like chimps, dolphins need to succeed within a complex social system. They face changing conditions that require complex cognitive abilities. Dolphin societies (also like chimps') are fission-fusion. There are shifting alliances among relatives and nonrelatives. There are even alliances of alliances in bottlenose dolphins, which is rarely seen beyond human societies. Dolphins are active social learners — they learn by observing the behavior of others, very much like us. Adults and young continue to show high levels of behavioral flexibility in the face of changing social and environmental scenarios. Dolphins learn much of their vocal and behavioral repertoires and variations within their social groups. You might say they share a culture: New vocal signals and behaviors are transmitted among members of a social group and from generation to generation. This is another trait that is shared between dolphin and primate species.

When two different species possess a similar behavior or body form as a result of being exposed to similar environmental circumstances, biologists call it evolutionary convergence. With dolphins and chimpanzees, with self-recognition and other higher cognitive abilities, we have a remarkable example of *cognitive convergence*. Dolphins and great apes last shared a common ancestor thirty million years before dinosaurs became extinct. Their brains' architectures are dramatically different. Yet their abilities have evolved similarly, for similar reasons.

The fact that we have evolutionary convergence to similar brain *functions* with differing brain *structures* is forcing cognitive biologists to rethink what underlies consciousness and the sense of self. Perhaps it isn't so important to discover the precise location in the brain of the sense of self. Perhaps it is an emergent property of any brain that has crossed some encephalization threshold or that contains the right number of neurons, has the right extent of connectivity, and includes the right organization of the connections. Whatever the answer, long-held assumptions about a revered function of the human brain have been shattered by watching dolphins in the mirror.

8

REFLECTIONS ON
DOLPHIN MINDS

AT MARINE WORLD, in Vallejo, we were lucky to be near the Lucasfilm company (creators of *Star Wars* and all that). One of the model makers and creature builders there, Mark Thorpe, became very interested in our dolphin work, and he often made exotic objects for Delphi and Pan's toy collection. But as is so often the case with kids (the dolphins were just three years old at this point), they were much more interested in mundane objects than in the fanciful creations from the *Star Wars* folk. A boat buoy in the shape of a fish, for example. It was physically robust, made of the same tough white plastic that's used in generic boat bumpers, which is a good thing for dolphin toys.

Delphi would grab the fish by its tail, swim directly to the bottom of the pool, and then release it. The fish would make its stately way to the surface, through sixteen feet of water, swaying from side to side as it went. The first time Delphi did this, he watched the fish's progress very carefully. He then zoomed to the surface, grabbed the fish again, took it to the bottom, and released it. This time Delphi swam up to the surface alongside the fish, swaying side to side, exactly mimicking the buoy's motion. It was like a pas de deux between a piece of plastic shaped like a fish and a marine mammal pretending to be a plastic fish. As I so often did, I gave enthusiastic encouragement from the side of the pool, shouting, "Good boy, Delph! Good boy!"

And as *he* so often did, Pan closely watched what Delphi was up to. Very soon Pan swam over, and a new game ensued. From then on, the boys took turns taking the fish to the bottom of the pool and accompanying it in a dance to the surface. This was such a dolphin-like episode: Delphi observing the fish's movements, then imitating them; Pan watching them both, then taking his turn, following exactly what Delphi had done; and then the two of them taking turns cooperatively. Dolphins learn through imitation in a social context, and this was a wonderful example. But it didn't end there.

One day not long after the fish game began, Delphi was on his way to the bottom of the pool, fish in his jaws. But before he got to the bottom, the fish slipped out of his grasp and prematurely started its return to the surface. I noticed that the fish wasn't doing its usual dance, no doubt because of its unusual orientation when it was released. But what happened next was surprising: for the first time, Delphi let out a big bubble of air that quickly morphed into a beautiful, silvery bubble ring. "Good boy, Delph," I shouted, and made a big circle movement with my finger, mimicking the shape of the bubble ring.

I had seen other dolphins blow bubble rings in the show pool, but neither Delphi nor Pan had done it, much less in the research pool. Delphi's bubble ring was pure serendipity, I'm sure of it. He had simply let out a burst of air on being surprised at the fish's unexpected release and unusual ascent.

"Good boy, Delphi!" I repeated, and again described a big circle with my finger. Delphi was at the surface by this point, watching me closely. He immediately swam back to the bottom, stationed himself in a deliberate manner, and blew another perfect bubble ring. Fast learner. The ever-vigilant Pan was watching, and he immediately came over to where I was standing at the pool's edge, swam to the bottom, positioned himself with deliberation, and blew a perfect bubble ring too. So began a period of a couple of weeks when the boys engaged in bubble-ring behavior, often when I was standing at the pool's edge, usually taking turns. As they alternated making bubbles, they would look up at me. Was it for my benefit? I had certainly lavished them with vocal praise and lots of attention. Again, so dolphinlike: they observe; see what works; see what gets attention; and then execute it perfectly.

I give this example of imitation and rapid learning as a lead-in to asking, Just how aware are dolphins? Every creature, no matter its cognitive endowment, is aware of itself in some sense, otherwise it would not be able to navigate the physical environment, find mates and food resources, and avoid predators. It's a matter of survival. What is it like to be in the mind of a dolphin in terms of self-awareness? How brightly does the spark of consciousness glow in these minds in the water? There is no way to answer this question directly, of course, but we can examine behaviors that reveal active minds, minds that are constantly searching, constantly exploring, constantly learning. The fish-and-bubble-ring story is one such behavior. But consider the behaviors mentioned in earlier chapters, along with some new examples:

The time-out episode: Recall young Circe, the first dolphin I worked with, in the small marine zoo at Port Barcares, near Perpignan, in southwest France. Circe had been observing my behavior, and she used it right back at me when I gave her a food she didn't like. This episode was, for me, a visceral experience as well as a scientific observation. I felt that crunch in the stomach, the shock of recognition. This wasn't just mimicking; it was communication.

Keyboard work and vocal mimicry: It's hard to imagine anything more alien from Delphi and Pan's natural environment than arbitrary symbols on a keyboard, and yet, without being schooled in the relationship between objects and symbols, they learned for themselves very rapidly through experience. And at the same time, they learned through experience the association of these symbols and objects with the arbitrary sounds we created.

Delphi and Pan quickly learned to imitate the arbitrary sounds, produced their own accurate renditions of them before they hit the appropriate symbols on the keyboard, and then spontaneously used them when playing with the corresponding objects. They never matched the sounds with the wrong symbol or object. Pan even seemed to use the ball whistle in a vocal exchange with Delphi, after which Delphi passed him the ball.

The other incident around the keyboard work that speaks of active minds involved Pan and his predilection for fish. Recall that on one occasion we decided to remove the fish symbol from the keyboard since Pan repeatedly activated it because he loved getting fish so much. But when he saw the keyboard with no fish symbol on it, he apparently assessed his own situation — his desire for fish; the keyboard situation — no fish symbol; and then figured out how to communicate his wishes to us, despite the hurdle he faced. He searched around the pool, found a remnant of fish, brought it to the keyboard, and touched a blank key with it, as if to say, I want fish! Creative problem solving.

Bubble-ring behavior: The aesthetics and easy inventiveness that take place around the bubble rings belie the cognitive abilities that bubble-ring production demands. There has to be the initial learning of how to make them, by chance (as in the case of Delphi) or by observation (in Pan's case); anticipatory planning and motivation; and an intuitive sense of the physics of what makes good rings and how they might be manipulated without destroying them. So here we have dolphins making arbitrary play objects appear out of thin air, planning and releasing the rings with exquisite timing.

Deception: Having another individual believe, based on one's own calculated behavior, that something is true when it isn't certainly demands high cognitive abilities; one has to manipulate the environment so that another individual is somehow fooled by an otherwise innocent action. True deception implies an individual knows the rules of the game and is manipulating them. It demands an appreciation of how the self's honest or dishonest act might be interpreted by the other. As you have seen, dolphins are quite capable of deception.

Mirror self-recognition: You saw in chapter 6 that Presley and Tab each definitively recognized the image in the mirror as self. And Delphi and Pan's behavior at the mirror was strongly suggestive of that ability as well. Recognition of self is, for obvious reasons, the strongest indicator that the glow of consciousness is more than a dim ember. But recall the urgency with which Presley tried to find the mark on his side, contorting himself in the cramped corner of the indoor pool after sham marking; remember his double take on first seeing his image as he casually swam by the mirror, as well as his horizontal swirl performance; think about the intensity with which he searched for a glimpse of self, looking into his own eye just inches from him in the mirror. Consider all these things, and it is not a stretch to put yourself in his mind and experience a consciousness that you recognize in yourself.

We *assume* consciousness in dolphins shines less brightly than it does in our minds. But this may be just an assumption. How would we know? How could we know? There is one thing we can be certain of, though, and that is that the *texture* of Presley's consciousness, the texture of all dolphins' consciousness, will be different from ours, just as Wittgenstein's lion inhabits a reality that is foreign to that of humans. One major reason: While humans are primarily visual creatures, dolphins experience their world primarily through sound, through their exquisite echolocation system. If you are a dog owner, or even simply a casual observer of a dog on a walk with its owner, then you are aware that a dog's experience of a walk is dramatically different from its owner's. The dog's world is constructed from a kaleidoscope of odors — a sensory realm that is mostly invisible to us — and visual images. In the same way, dolphins can perceive their immediate environment via their exquisitely detailed natural sonar. They receive acoustic pictures built from the returning echoes of the clicks that reflect off environmental features. The upper ranges of their sonar clicks are ultrasonic to us, beyond our perception. How they integrate and interpret the returning echoes of the clicks they've produced is still a mystery. No human-made sonar comes close to this natural technology.

Lou Herman has been a professor of psychology at the University of Hawaii for four decades, and in 1970 he founded the Kewalo Basin Marine Mammal Laboratory in Honolulu to study perception, cognition, and communication in bottlenose dolphins. A little more than two decades later, he and his colleague Adam Pack founded the Dolphin Institute, which added conservation and public education to other research interests. Lou directed what is likely the most longitudinal cognitive research program with dolphins; it began in 2004 and has resulted in many groundbreaking experiments on dolphin perception and intelligence.

Early on, Lou Herman bravely decided to investigate dolphins' language competence. I say *bravely* because the investigation of animal language has a long history of contentiousness, thanks in part to John Lilly. Whereas Lilly had dreamed of humans' one day having two-way conversations with dolphins in spoken English, Herman concentrated only on dolphins' ability to comprehend, not produce, language. And he didn't use spoken words, as Lilly had wanted to. Instead, he taught one dolphin, Akeakamai, an arbitrary system of hand gestures with a simple grammar, akin to American Sign Language, as some ape-language researchers had done. A second dolphin, Phoenix, learned another arbitrary system, dolphin-like whistles created by Herman and his colleagues.

The Kewalo dolphins quickly grasped not only the meaning of the symbols (for hoop, ball, fetch, and so on, which can be thought of as the semantics of the language) but also the meaning of the word order (the syntax). For instance, when one of the trainers gave Akeakamai the symbols for hoop-ball-fetch — in that order — she would push the ball to the hoop. But when she was given the instruction ball-hoop-fetch, she took the hoop to the ball. We all know that word order in sentences is critical to comprehension, and we respond to it without thinking. Herman put it this way: "Syntax is what tells us that a venetian blind is not a blind Venetian." Both Akeakamai and Phoenix were almost flawless in their response to the meaning of word order in the simple, and sometimes not so simple, sentences Herman gave them. For instance, they knew that "take the pipe to the basket on the right" meant that the pipe should finish up in the basket on the *right*, not the one of the *left*. Lou's dolphins proved themselves at least up to par with any chimp in their grasp of semantics and syntax.

Akeakamai and Phoenix also mastered other cognitive tests. For example, a trainer would sometimes ask one of the dolphins to fetch a novel object and do something with it, and the trainer would use a symbol for the object that the dolphin had not yet learned. The dolphin had no difficulty figuring out that the symbol must refer to the only object in the pool that it hadn't seen before. Akeakamai also understood the rather abstract concept of presence and absence, and used Yes and No pedals to respond to questions. For instance, when she was asked, in the form of the symbols that meant "ball" and "question," whether there was a ball in the pool, she would press a Yes pedal if there was a ball there and a No pedal if there wasn't. Sounds simple, doesn't it? But dealing with an abstract concept, as absence is, demands more brainpower than dealing with the concrete.

Akeakamai spontaneously and on her own took this little exercise further. Akeakamai knew that the construction Frisbee-hoopin meant "put the hoop on the Frisbee." One day, she was told to do this but there was no Frisbee in the pool, so she got hold of the hoop, went over to the pedals, and put it on the No. Another time, when the object to be moved, the hoop, was absent, she simply went to the pedals and pressed No with her rostrum. She had found a way to respond correctly. She had worked out how to say "There's no Frisbee" and "There's no hoop." She'd devised these responses to impossible situations on her own, with no training. That's a reflection of dolphin mind: problem solving and communicating.

For me, one particular study that Herman and Pack conducted

was remarkable and extremely revealing about dolphin sensitivities — and perhaps about their own form of communication. Akeakamai not only responded to the meaning of gestures with impressive accuracy but also showed herself to be superbly sensitive to the physical form of the gestures that carried their meaning. Ordinarily, a trainer made gestures standing at the side of the pool, moving an arm from a bent position to straight out in front, moving it across the body, and so on. In this particular experiment, the goal was to determine how dolphins interpreted signals that were visually degraded. Rather than generating gestures from the normal poolside positions, the trainer gave all signals from an underwater viewing window. The first stage of degrading the signal was to have each trainer dressed in black, with his arms in white. Akeakamai would therefore see just the arms as the trainer gestured. She was able to respond just as well as she normally did. The next stage was for the trainer to wear a black body suit, with only the hands in white. Akeakamai now saw only the hands, and she still had no problem. The last stage had the trainer totally in black and holding a short black stick in each hand, a white Ping-Pong ball at the end of each stick. Akeakamai still aced it, which astonished Herman and his colleagues because, given the minimal information, they themselves had tremendous difficulty identifying which gesture was which.

This experiment was done quite some time ago, but when I saw Lou Herman present the videos at a conference I flashed on something I had experienced when swimming with dolphins. On occasion I would swim with Circe, Terry, Pan, and Delphi. While behind them, I noticed small white areas on the leading or back edges of their pectoral fins or flukes. From the front I noticed the tips of their rostrums were lighter as well. I used these markers as directional cues so I could maintain cohesion and velocity with the group and sense which way they would move next. Birds use visual cues in similar ways in maintaining flock cohesion. Perhaps the dolphins themselves used these white-tipped cues? Recently I was swimming with dolphins during a research trip with Daisy Kaplan, a doctoral student in my lab, off Bimini, in the Caribbean, and I wondered if they were using the same visual cues that I used with them.* Perhaps you have seen films of groups of dolphins swimming in unison, twisting and turning, diving and leaping, as if they were perfectly executing a piece of choreography. How do they do it? How do they avoid bumping into one another? Visual monitoring of neighbors must be a part of the system, though probably not all of it. So what are they actually monitoring? It was a wow! moment for me when I saw Lou Herman's video. Dolphins have exquisite sensitivity to minimal information, and they can accurately act on it.

Herman also taught the dolphins the symbol meaning "mimic," which told them to copy the body movements a trainer performed at the edge of the pool. The act of mimicry sounds simple, but it means that one individual must model its behavior based on that of another. And when the other is a different species with completely different anatomy, an extra cognitive step is involved: matching analogous parts of the observer's anatomy to the demonstrator's. Herman's dolphins had no problem copying a whole-body pirouette, relating whole body to whole body. No surprise there. They were also adept at copying a trainer bending backward from the waist with her hands by the sides of her head. Still not too challenging. When the trainer shook a leg, that should have presented a problem. But the dolphins shook their tails! We've already seen how accomplished dolphins are at vocal mimicry (more so than any animal other than certain birds and humans). This, combined with behavioral mimicry, makes dolphins the only species, aside from humans, that are capable of both forms of mimicry.

Dolphins appear to be performing a dance as they swim synchronously in groups in the ocean. Who is devising the dance? And how is it communicated among the group? Lou Herman decided

*For video of wild dolphin behavior in Bimini, see www.hmhbooks.com/dolphinmirror# wild.

to dabble a little bit in this performance arena. He taught Akeakamai and Phoenix a symbol meaning "create," which told them they should go off and do something novel. (Karen Pryor had done a study earlier showing that dolphins could learn the concept of "do something novel and creative"!) He has a film segment that showed a trainer giving Akeakamai and Phoenix this command at the side of the pool. The two dolphins swam to the center of the pool and circled underwater for a few seconds. Then they dramatically broke through the surface of the water, spinning clockwise and erect, heads up, and squirted water from their mouths, all in such perfect unison it would have made Olympic synchronized swimmers proud. "None of this was trained," Lou Herman said, "and it looks to us absolutely mysterious. We don't know how they do it."[1] These active, intelligent minds understand *novel* and *create,* and they can do it in teams.

Lou Herman and his colleagues have produced many findings over the years that further indicate that dolphins are the cognitively complex creatures we intuitively feel they are. I've given only a glimpse of his work here, and I wish I had space for more. But I do want to add just one last finding of Lou Herman's. It's an experiment I wish I'd done.

Dolphins' sonar systems are, quite properly, regarded with something approaching awe by technical and nontechnical people alike. They are the subject of both impressive data and questionable claims. In the first category, impressive data, the U.S. Navy has determined that a dolphin can accurately locate a three-inch sphere from a distance of four hundred yards. The second category, questionable claims, arises from the fact that the sound waves that dolphins produce are able to penetrate biological tissue. In principle, therefore, dolphins should be able to examine one another's internal organs at a glance. What they "see" and what use they make of such information is a mystery, but there are stories that imply dolphins sometimes know more than we think possible. I have heard numerous accounts of people who claim that when they swam in the wild with dolphins, injured parts of their bodies were scanned by the dolphins and manipulated. This may be the case, but it is certainly possible that one moves an injured limb differently, which could capture the attention of a dolphin and lead it to investigate. I've also heard numerous accounts of pregnant women who claim that dolphins used sonar on their abdomens. It's not surprising to think that dolphin sonar could reveal a pregnancy. After all, our less sophisticated ultrasound technology performs this task. I now conduct observations and cognitive studies with a social group of dolphins at the National Aquarium in Baltimore. The social group is composed of several generations of females and a few male offspring. We have observed an older female scan the abdominal area of her pregnant offspring. It would certainly confer an advantage to the individual and the social group to know who is expecting. Whether they are really detecting a pregnancy and understanding the ramifications, however, is far from clear.

What isn't in doubt, however, is the insight into the dolphin mind that Lou Herman's beautiful experiment gives us about what dolphins "see" when they echolocate. Just as it is hard for vision-oriented creatures to imagine what a dog "sees" when it is creating an olfactory picture of its world, so it challenges our imagination to picture a dolphin's mind when it creates an acoustic picture of the world. But Herman and his colleagues provide compelling evidence that dolphins' sound-based perceptual systems produce something akin to visual images.

Lou collaborated with Adam Pack and Matthias Hoffman-Kuhnt on this classic experiment with a seven-year-old female named Elele. Elele was an eager student in whatever she did, and she loved to be right, which she was most of the time. When a trainer gave her the "hooray" signal, both arms stretched up vertically, Elele would leap into the air, squeaking and clicking with delight. In this experiment, Elele had a lot to squeak about, because most of the time she was near perfect in her performance.

The experiment was simple in concept but demanded careful

execution. The question was this: Would Elele be able to visually identify an object after she had sonared it? In other words, does the sonar echo translate into a visual image in a dolphin's mind?

Lou Herman and his colleagues had Elele examine the contents of a box that was transparent to sonar but visually opaque. Inside the box was a complexly shaped object constructed from PVC piping, measuring about ten inches in size. Elele's task was to visually identify the object she'd "seen" by way of sonar. It involved her discriminating between two very similarly shaped objects and picking the right one. Elele had never seen the objects before, so there was no element of recognizing something familiar. She had to form a clear image of a complex novel object based only on sonar and prove that she'd seen the object by choosing it over something very similar. She got it right almost 100 percent of the time. And she did it almost instantly. We still don't know exactly how the sonar-produced image is manifested in the dolphin's brain, but we can now say that it is not merely recorded in sound. Perhaps it is some kind of holograph? Only dolphins know.

Elele showed herself to be equally skilled in the reverse direction, going from visual image to sound image: she had no difficulty identifying an object with sonar that she had seen visually just once. This was a very important study. It would be fascinating to repeat it while rotating the objects slightly in the visual-identification phase to see if dolphins can mentally rotate something. I bet they can.

❧

Spock and Shiloh were inseparable, always swimming together, always playing together, always resting together. They were a major item in the show pool at Marine World in Redwood City. Everybody loved this devoted pair. Spock was the male dolphin who had apparently tricked Jim Mullen into giving him multiple rewards for cleaning up the pool by bringing him multiple scraps of paper torn from one large piece. And Shiloh was the first dolphin I saw producing bubble rings, soon after I arrived at the facility in 1981. Then, quite suddenly, Spock became ill, and died. I'd known him for only a few years and hadn't worked closely with him in the way I worked with Terry and Circe and their boys, but I was stricken, as were Jim Mullen and the other trainers. And we were not alone.

Shiloh was right by Spock's side as we lifted his inert body out of the pool. She looked bewildered and bereft. She wouldn't eat. She no longer swam around the pool with joie de vivre. She spent a lot of time simply lying on the bottom, deeply lethargic, stirring only when she had to surface to breathe. The other dolphins became very solicitous of her, swimming up to her side, apparently trying to encourage her to swim with them. But the grieving Shiloh would not be consoled. She continued like this for some time, and we became quite concerned for her health. Thankfully, after several days, like a human emerging slowly from mourning, Shiloh began to eat and socialize with the other dolphins. Oddly familiar. A pattern that connects.

There are a number of words in the preceding paragraph that, strictly speaking, should be in quotation marks. *Bewildered* and *bereft,* for instance; and *solicitous* and *grieving.* These words are inferences of the states of mind of animals, not states that we know for certain. With known human behavior as our model, we thought Shiloh looked *as if* she were bewildered and bereft; her buddies behaved *as if* they were being solicitous of Shiloh, concerned about her well-being. If someone you knew behaved after the death of her mate as Shiloh behaved when Spock died, you would know she was grieving, and you would have a very good idea of how she felt, especially if you yourself had experienced such a loss. Shiloh looked *as if* she were grieving. Does that mean that she was experiencing something like the raw, searing, heart-rending emotions we associate with grieving? How can we know?

Consider this. Back in the 1960s there were several reports of dolphins in aquariums displaying caregiving, or epimeletic, behavior with other dolphins. Occasionally, an individual dolphin in a pool becomes sick, loses strength, and is in danger of drowning because it is unable to swim to the surface to breathe. Often when this happens, other dolphins come to its aid; they stay at the sides of the ailing individual, holding it up so that it doesn't sink and drown. It looks as if the rescuers recognize that the individual is in trouble and care enough about it to keep it from drowning, often forgoing feeding for a long time while on the rescue mission. Is this a hard-wired response to a flailing individual, or do the dolphins really understand the situation and know what they are doing? Is their caregiving behavior evidence that they understand the plight of another?

Another situation. From time to time we do routine health checks on the dolphins. We teach each individual certain behaviors that help us carry out various procedures, such as lying belly-up at the water's surface so we can inspect its underside or do an ultrasound on an expectant mom; keeping its mouth open so we can examine teeth, gums, and throat; keeping its blowhole open so we can take a swab for a culture. When we did this with Gordo, Terry and Circe would abandon their carefree swimming around the pool and stand vigil until we released Gordo back to them.

When Gordo was actually sick and needed more invasive medical attention, we lowered the water level in the pool, put a foam mat under him, and supported him in a sling. Under these circumstances, Terry and Circe became seriously attentive and seemingly stressed, so much so that we had to take vigorous actions to keep them at bay. They acted *as if* they were truly concerned about the welfare of their buddy; they appeared to be agitated, and there was a lot of whistling back and forth. They acted *as if* they cared. There is nothing unique about Terry and Circe in this respect. Scientists and trainers who have worked with dolphins routinely see such behaviors in individuals, *as if* the dolphins feel the distress of a buddy in a difficult situation and experience a need to help. It may be the dolphin equivalent of Bill Clinton's saying, "I feel your pain."

The behaviors I am describing here are what we would call

empathy or, more simply, caregiving, and dolphin lore is replete with such tales. One of the earliest examples to find its way into the scientific literature was reported in 1966 by M. C. Caldwell and D. K. Caldwell.[2] This observation was of a very unfortunate incident that occurred off the coast of Florida in late October 1954. In the 1950s there was little concern or protection for dolphins, and we were living in the dark ages in our understanding of animal minds. In what can only be described as an act of pure ignorance, at an exhibition for a public aquarium someone set off a stick of dynamite near a group of bottlenose dolphins. The apparently lifeless body of a victim of the explosion rose to the surface, listing awkwardly. Two dolphins immediately came to its aid, one on each side of the stunned individual, supporting it so that its blowhole was above the water. Being underwater themselves, the supporting animals were unable to breathe, and after a few minutes they had to leave their post, only to be immediately replaced by two other individuals. This relay continued until the stunned animal regained consciousness, whereupon it quickly swam away. The rest of the dolphin group had remained nearby, rather than fleeing, and now accompanied the recovered individual. The scientists who reported the incident said, "There is no doubt in our minds that the cooperative assistance displayed for their own species was real and deliberate."[3]

Real and deliberate is the key phrase here. It implies the behavior is the product of active intelligent minds, not a blind, hard-wired pattern that merely appears to be a conscious act. Critics argue that caregiving of this sort is just an extension of a hard-wired protective behavior of dolphin mothers with their newborns, who often flail helplessly and sometimes need to be pushed to the surface to prevent them from drowning. But even this protection of newborns is anything but hard-wired. I have seen several newborn dolphins on the verge of drowning, their inexperienced mothers not having a clue what to do, as happened when Circe gave birth to Delphi, and Terry had to intervene.

Moreover, if this kind of supporting behavior were hard-wired, we would see it on *every* occasion when an individual is in trouble. Ken Norris tells of an incident in which a pilot whale had been harpooned and its two companions did the opposite of the usual caregiving behavior. Initially they stayed very close to their mortally wounded companion, one on either side, supporting it in the usual way, as it was being reeled in. Yet as the dead animal got close to the ship, the two Samaritans abruptly changed their behavior, leaping onto the lifeless body and pushing it down and away from the ship.[4] This is not an isolated incident; similar stories have been reported elsewhere. Rather than being hard-wired, therefore, the response of dolphins and whales to an injured or killed companion shows flexibility encompassing behaviors that are appropriate to particular situations. This flexibility implies a conscious awareness of the nature of the prevailing circumstances, and a conscious decision as to what action should be taken.

Dolphins and whales are well known for sticking by a hurt or mortally wounded companion, a behavior we might regard as admirable in human terms, displaying as it does a sense of loyalty and care. But it also makes them easier to hunt, a trait that whalers have exploited through the ages. A mid-nineteenth-century report is one of the earliest published observations of such behavior. Whalers spotted a pod of Pacific pilot whales off the coast of Panama and managed to harpoon one. "The school hovered around their injured schoolmate and refused to depart for some time," explained a later review of the behavior, "even after the dead animal had been hauled aboard ship."[5] Hanging around the scene of loss leaves the pod vulnerable to further killing, which was exactly what whalers often did.

Dolphins displayed the same all-for-one, one-for-all behavior in the infamous hunt drives at Taiji, Japan. The dolphins are driven toward shore and trapped behind nets in the bay; the dolphins have the physical ability to leap over the nets and escape, but their desire to stick together is apparently too strong, ironically leading to the wholesale slaughter of entire social groups, including mothers and young.

Earlier I mentioned stories from ancient times of dolphins showing caregiving not only to their own kind but also to other species, most notably humans. Some of these tales may be just that, stories based on a sliver of truth that became embellished in the telling through the ages. But contemporary accounts are plentiful enough to allow us to suspend some disbelief. On one high-profile occasion, Elian Gonzalez, a Cuban six-year-old who was fleeing with his mother to the United States in 2000, survived for two days in the Caribbean lying on an inner tube after his boat sank, his mother having gone down with it. The two fishermen who plucked Elian from the sea said there were dolphins circling the boy on his tube. And Elian himself told reporters that dolphins surrounded him and would push him back up onto the mini raft when he was losing strength and slipping off. I have an acquaintance who assisted young Elian in the days following his rescue, and the boy claimed that the only time he felt safe was when the dolphins appeared. Similar sentiments have been expressed by many others who have been on the receiving end of dolphin rescues.

I have a tale of a dolphin rescue myself, albeit a vicarious one. I was a guest on a radio show shortly after the Asian tsunami of December 2004, and I was asked to speak about whether animals have special sensory systems that warn of impending disasters. A man called in with the following story. He and his family were vacationing in Indonesia and were in a small boat far offshore, fishing. All of a sudden they noticed fins circling their small craft. They thought they were about to be attacked by sharks. But it turned out that the fins belonged to dolphins, which proceeded to push the boat to shore. Then the tsunami hit. "These animals saved me and my family," the man told me. "I am convinced of it." Not long afterward, I got a phone call from a woman in Greece who said that dolphins had saved her life too. "I go swimming in the sea all the time," she told me, "and I often see dolphins, but they never come near me. But on this one occasion I got into trouble and thought I was going to drown. Then I felt a nudge, and I was moving rapidly toward the shore; this dolphin was pushing me. The dolphin saved my life."

Here is a story of an incident off the coast of Venezuela near Isla de Margarita, a place where it is said that people have a special relationship with sea creatures. Tony Salazar was on a sailboat with his brother, taking part in the first race of the South Caribbean Ocean Regatta, in June 1997. With brisk winds and a choppy sea tossing the boat around, Salazar fell overboard while trying to execute a maneuver with the spinnaker. Because the boat was moving so fast, by the time the crew turned it around Salazar was nowhere to be seen. Salazar screamed and waved as he watched the boat disappear into the distance. He thought he was doomed. After about half an hour, he realized he was surrounded by dolphins, which was a great relief to him, as he knew that dolphins are natural enemies of sharks, which were common in those waters.

Meanwhile, the powerboat *Gala* was searching for Salazar in a zigzag pattern. The crew noticed a pair of dolphins approach and then turn and swim away. This happened several times, and the *Gala*'s captain had a hunch that the dolphins were trying to tell him something. He turned the *Gala* to follow the dolphins, even though they had already searched in that vicinity. After a short while the crew spotted Salazar and hauled him out of the waters, exhausted, numb, and cramped after this two-hour ordeal. It was Salazar's fiftieth birthday, and he really had something to celebrate.[6]

A decade earlier, a story of dolphins rescuing not one but three sailors in trouble was reported. Peter Stock, Terry MacDonald, and Roger Hilligan had been sailing a mile from the mouth of the Great Kei River on South Africa's east coast in rough weather when their yacht capsized. The three men managed to stay close to the boat and struggled to right it, but they felt themselves being dragged out to sea, toward waters where they had seen sharks the previous day.

Understandably, the shipwrecked sailors were apprehensive. But then dolphins appeared on the scene. "It was only then that I didn't mind falling into the water," Stock told the *Johannesburg Star*. "I felt safe." Stock said that the dolphins stayed with the boat as they clambered back aboard, and even steered it to safety, away from rocks and toward shore. "As soon as we were all safely ashore they disappeared," Stock said. "They gave us a feeling of security and spurred us into action."

In the book *Beautiful Minds*, coauthor Maddalena Bearzi tells an extraordinary story of the efforts of a group of dolphins to save the life of a young woman off the coast of Malibu, California. Bearzi was studying the foraging behavior of the group of nine dolphins as they encircled a large school of sardines just off the Malibu pier. Soon after the dolphins started feeding, one of them abruptly left the circle and swam away at high speed. Almost immediately the other eight dolphins abandoned their feeding and followed their companion. "This was an odd behavior for my metropolitan dolphins," Bearzi observed. "To abruptly stop feeding and take off in an unrelated direction was rather peculiar."[7]

Curious, Bearzi set off in her powerboat in pursuit of the dolphins. About three miles offshore the dolphins came to a sudden stop and formed themselves into a circle, for no reason that was obvious to Bearzi. "That's when one of my assistants spotted an inert human body with long, blond hair floating in the center of the dolphin ring," wrote Bearzi.[8] She maneuvered the boat toward the unconscious woman, and the crew hauled her aboard. The dolphins swam away. The young woman was hypothermic and would have died had she been in the water much longer. This is an extraordinary tale. How did the first dolphin know that someone three miles distant was facing death? And why would the group be motivated to save her?

I have a collection of ancient myths and modern tales from newspapers about dolphins coming to the rescue of humans. I have a drawer full of them. There must be some truth to these stories.

(Although if dolphins ever did the opposite, pushed a person away from safety to his death, how would we ever hear about it?)

Why does the impulse for caregiving exist among dolphins and whales, if indeed these stories really do imply the behavior of active thinking minds? Gordon Gallup was the first to point out that behaviors that can be described as caregiving occur only in animals that have large, complex brains and are able to pass the mirror-self-recognition test. Membership in that club is small: humans, great apes, dolphins, elephants, and, as recently discovered, magpies. "Organisms that are aware of themselves are in a unique position to use their experience as a means of modeling the experience of others,"[9] he said. This is not limited to having insight into another individual's practical reactions, but also into their mental states, their emotions.

The primatologist (and my colleague and collaborator) Frans de Waal, author of *The Age of Empathy*, said that in chimpanzee societies, instances of one individual actively consoling another, often after a fight, are quite common. "A victim of aggression, who not long ago had to run for her life, or scream to recruit support, now sits alone, pouting, licking an injury, or looking dejected," wrote de Waal. "She perks up when a bystander comes over to her to give her a hug, groom her, or carefully inspect her injury. Consolation can be quite emotional, with both chimps literally screaming in each other's arms."[10] Does the bystander go to the victim of the fight and give comfort because she knows that under those same circumstances she, too, would be emotionally hurting and would like to be comforted? Do Terry and Circe stand vigil over Gordo when he is having a medical procedure because they know that they would feel distressed under those circumstances and would want to be comforted?

It is a difficult topic for humans to discuss dispassionately because, being the super-empathetic creatures that we are, it is impossible for any one of us to see or even imagine another distressed and not feel an emotional tug in the stomach and an urge to give comfort. But it remains true that the existence of a theory of mind in a species does not *automatically* lead to the experience or expression of empathy and caregiving in one individual for another. Mind reading could, in principle, be an experience entirely free of emotional identification. Psychopaths can "read" other people — they just don't care a whit about them. Some further cognitive step needs to be taken to go from emotionless mind reading to identification with another's psychological state to the urge to ameliorate suffering or distress.

I suspect dolphins and whales, with large, complex brains living in complex societies, may use their smarts to read the psychological states of other individuals as a tool. Some biologists balk at this idea, most notably Daniel Povinelli. "Gallup speculates that the capacity for self-recognition may indicate that chimpanzees are aware of their own internal, psychological states and understand that other individuals possess such states as well," he wrote. "I have come to doubt this high-level interpretation of chimpanzees' reactions to seeing themselves in mirrors."[11] Povinelli argued that the brains of great apes are exquisitely wired to monitor the position of every part of their bodies, at all times. He describes this as chimpanzees' "kinesthetic self-concept," an ability that allows great apes to navigate deftly and safely in a hazardous environment. To Povinelli, humans can only be certain about their own mental states and can infer things about the mental states of others. "Other species, including chimpanzees, may simply be incapable of reasoning about mental states — no matter how much we insist they do."[12]

Danny Povinelli could be right, of course, but I believe his is an extreme position. As I have said on more than one occasion, when one spends time in the presence of dolphins and becomes highly attuned to them, as I have, or with chimpanzees, as Sue Savage Rumbaugh has, or with African Grey parrots, as Irene Pepperberg has, one forms the strong impression that there is "somebody" there, the presence of an intelligence that goes beyond a neural machine monitoring the physical actions of the body. While I am aware of the snares of anthropomorphism, I am also aware of the patterns that connect us, and when I am with a dolphin, I feel in my gut that I am in the presence of mind, not just body.

There seemed to be mind at work in a remarkable episode that involved the rescue of another humpback whale, this time near the Farallon Islands, about eighteen miles off the coast of San Francisco. In mid-December 2005, the animal, which measured some fifty feet long and weighed around fifty tons, became desperately entangled in hundreds of feet of crab-pot ropes and multiple weights. One line was in the whale's mouth. Crab fishermen spotted the whale early on a Sunday morning, and by that afternoon a rescue team was at the spot to attempt what would be an extremely hazardous venture. Divers had to swim around and under the whale in order to cut the ropes free or it would sink and drown. Yet with one flip of its massive tail, a humpback can kill a man. "I was the first diver in the water, and my heart sank when I saw all the lines wrapped around it," said James Moskito, who worked for a cage-diving outfit and led the rescue team. "I really didn't think we were going to be able to save it."

Moskito and three other divers spent about an hour cutting the ropes with a special knife. Some lines were so tight that the divers had to dig deep into the whale's flesh. During the entire ordeal the whale remained calm, as if it knew that the men were there to help it. "When I was cutting the line going through the mouth, its eye was there winking at me, watching me," said Moskito. "It was an epic moment in my life." When it was finally free, instead of immediately swimming away, the whale made small circles around the men, and then nudged each of them in turn. "It felt to me like it was thanking us, knowing that it was free and that we had helped it," said Moskito. "It stopped about a foot away from me, pushed me around a little bit and had some fun." Mick Menigoz, another of the divers, said, "I don't know for sure what it was thinking, but it's something I will always remember."[13]

I fully empathize with the emotional impact that this close

encounter had on those men, especially experiencing the lingering eye contact. It brings to mind that moment two decades earlier when Humphrey, having mistakenly started to swim north from San Francisco Bay, returned to our flotilla, came to my boat, bellied up to the side, and gazed at us for several long seconds as we looked down at him. Like the divers, I didn't know what Humphrey was thinking, but it was hard for me to believe that his eye was a window into a blank mind.

9

INTO THE COVE

I FELT A POLITE tap on my shoulder as I was peering at the poster display, my own, for the hundredth time, wondering whether people would be able to see the real story behind it. I turned and saw a tall, deeply tanned, silver-haired man next to me, his steel blue eyes holding me in a determined gaze. "Excuse me, but are you Diana Reiss?" I nodded. He was a total stranger to me, and he had a strong presence about him. There are a lot of big, bronzed, outdoorsy guys in my line of work, so this inquisitive stranger was nothing out of the ordinary. It was late in the afternoon, my session was coming to an end, I was tired, and I was looking forward to meeting up with some colleagues for a quiet dinner. I was unprepared for what the stranger said next: "I am interested in doing a film about the environment that will really make a difference and I have access to someone who has the monetary equivalent of a shah to do it. Someone told me I should talk to you."

It was the middle of December in 2005, and I was attending the sixteenth biennial conference of the Society for Marine Mammalogy (SMM). These events are the go-to venues for scientists engaged in research on marine mammals, such as dolphins, whales, seals, sea lions, walruses, sea otters, and so on. That year it was being held at the Manchester Grand Hyatt in San Diego, and there were more than two thousand participants, including five hundred students, thirty-eight companies and organizations displaying their wares in various ballrooms, more than three hundred oral presentations, and almost a thousand poster slots. It was, as this genre of scientific gatherings usually is, a zoo.

Five years prior to the San Diego meeting I had by chance become interested in what can best be described as distress calls, the kind made when dolphins are under duress, such as in pain or severe stress. On a spectrograph, the distress call looks like a short rising whistle followed by a longer falling whistle. It sounds to me like a falling bomb. John Lilly was the first to recognize distress calls, in the mid-1950s, and René-Guy Busnel, my professor in France early in my career, had suggested that bottlenose and other dolphins and whales produce distress calls when injured or in pain; for example, when harpooned. But the topic wasn't prominent in anyone's mind when I came across it through happenstance in the year 2000.

In the late fall of that year, a young female dolphin, which we named Mara, had become stranded along with an older female presumed to be her mother in the Shrewsbury River in New Jersey. Officials at the National Marine Fisheries Service had been monitoring the dolphins for weeks and decided they had to be rescued before winter came and the river iced over. The older female had internal injuries, probably from a collision with a boat, and other severe medical problems, and unfortunately, she died during the rescue attempt. Mara, however, was carefully transported to the National Aquarium in Baltimore by the aquarium's marine mammal rescue program. I was a visiting research scientist there at the time. Some rescue-team members told me that prior to the death of the older female, the two dolphins were exchanging a distinctive odd whistle that sounded like a falling bomb. I raced to the aquarium the next day to record Mara's vocalizations. She was medically isolated in the aquarium's rescue pools and ministered to day and night by the veterinarians and marine mammal rescue staff. She looked like a healthy, albeit lonely, little dolphin, but people were always there with her and provided her with a variety of objects to play with while the rescue team waited for the medication they'd given her to work. Sadly, after a short period, Mara died too, but not before I had had the opportunity to record whistles of the sort that Lilly and Busnel had identified. I was puzzled about why Mara had been producing distress calls so persistently, but the mystery was solved with the autopsy: she had fibrous adhesions throughout her internal organs, the same medical condition from which her mother had suffered; and stomach acid had leaked into her body cavity. The poor dolphin must have experienced major pain. Dolphins, like many other animals, often mask their pain. This is highly adaptive for many species because predators seek out weak, ill, or injured individuals. Only her whistles signified her suffering, though we didn't realize it.

A light bulb lit up for me. I had been searching for a Solomon's ring, a decoding cipher to understand how dolphins use their whistles to communicate with one another. Now I realized that if we could confirm that dolphins produce these falling-bomb calls when they are in physical duress and at no other time, then we would have a terrific veterinary and welfare tool; if an individual produced these particular whistles, we would know that it was in distress, and we could then try to ameliorate whatever was causing it. Two years later I gave a talk on my observations of Mara. The principal reason I had gone to the San Diego meeting in 2005 was to introduce some even newer data on distress calls. My poster presentation included audio of the calls and a sonogram, or sound picture, of them. I also had a photograph of one of the dolphins that had been producing the calls.

Poster sessions at these extravagant scientific gatherings give researchers an opportunity to display new research and research in progress, often prior to publication. Huge ballrooms are divided into multiple narrow passageways by temporary walls on which three-foot-by-four-foot displays are hung, some hurriedly put together, others the products of months of careful construction. I actually don't care for poster sessions. As a presenter, I feel like a vendor, hawking my wares, desperate to find a sympathetic ear. I much prefer to present a spoken talk to an audience rather than doing a poster presentation, and I generally do not submit papers for that purpose. Until the San Diego meeting.

The reason I decided to go against my rule was not so I could talk about the use of distress calls in veterinary or aquarium practice only, although my recent work was focused on that idea. My motivation was different, something that compelled me to step out of my usual role as a pure scientist and become a scientist-advocate. It was not an easy shift. There is great pressure in academia to remain pure and "above the fray." Many scientists believe advocacy should be left to others, to those who work with environmental agencies, political organizations, and other NGOs with a mission — to those who need not be concerned that their scientific reputations and credibility might be besmirched by straying into politics. But something sinister and disturbing pushed me past this objection.

About four years before the San Diego meeting, not long after I had pinpointed dolphin distress calls with Mara, I learned of a practice that had been going on for many years, a practice about which most people were unaware, even in Japan, where it was carried out. In a small coastal village, Taiji, in the southern part of the Japanese archipelago, there were yearly drives of hundreds of dolphins at a time; they were trapped in a blind cove, where they were brutally slaughtered by fishermen. By the end of each season, more than two thousand dolphins have been killed in this gruesome manner. I have a short video clip of one of these types of killings, and I always warn people who want to see it that the images are extremely disturbing. This gory video clip, which had been surreptitiously recorded, also had an audio track that carried distress calls of the kind produced by Mara before she died.

My stomach turns every time I watch the killing and hear the calls. Every time. Every time. My three decades of work with dolphins has taught me that dolphins are not biological automatons, devoid of feeling, emotion, or an awareness of themselves and what is happening around them. Bottlenose dolphins are sentient, highly social, and highly intelligent animals. The annual slaughter of bottlenose dolphins and other dolphin and whale species is therefore inhumane beyond words. For me, it is unjustifiable from any perspective. In 2001 I committed myself to doing whatever I could as a scientist-advocate to stop the killing. It became my mission, running parallel with my mainstream scientific research.

The 2005 San Diego meeting was the first time I had gone into the conventional scientific arena openly promoting this cause. The Society for Marine Mammalogy, like many scientific societies, discourages people from using its gatherings for anything but scientific ends. I had, however, managed to persuade society officials that the dolphin drives were such an egregious affront to marine mammal integrity and dignity, they should allow me to seek signatures to a petition calling for its termination. I had to agree that I wouldn't aggressively pursue people in the corridors to corral signatures. With help from colleagues and graduate students, we set up a demonstration, including a video of the drives, in a side room and invited people to sign the petition if they felt so moved.

We were seeking signatures from our colleagues, an international group of marine mammal scientists, and I wanted to make sure we had representation from both younger scientists and established researchers. My colleague Lori Marino worked with me on the petition, and we asked our colleague Sam Ridgway to help us draft the scientists' statement we were asking participants to endorse. Sam Ridgway was the world's preeminent dolphin veterinarian; he literally wrote the book on dolphin biology and physiology. Sam was an old friend as well as a colleague, and for political reasons, I didn't want the petition statement to be from the pens of women only. The long statement addressed to the government of Japan urged it to "lead the way and take action in stopping the inhumane treatment and killing of these highly sentient mammals." It turned out to be easy to get people to sign. About three hundred scientists did so during those few short days of the conference. We also posted a second public petition open to all on our website (www.actfordolphins.org), which I'd started with Paul Boyle, then the director of the New York Aquarium, and Lori. Within a year, the number of signatures had climbed to more than a hundred and twenty-five thousand.

Meanwhile, in the earnest academic intensity that was the poster session, my presentation was something of a paradox. It showed a photograph of a small cove and fishing boats, and another photograph of a single dolphin, its blood-covered head plaintively raised above the water. But then there was a spectrograph of an audio recording of dolphins being driven into the killing cove; it showed the type of whistles, the number of whistles, and an acoustic analysis of them. All very sterile and devoid of emotion.

It was something of a schizophrenic experience for me, this division between the scientist coolly amassing evidence and the scientist-advocate desperately trying to find ways of changing the world. Many of my colleagues know me as a passionate person with a determination to increase global protection for dolphins and whales. My fervor to stop the killing at Taiji was even more than usually high-octane. So when the silver-haired stranger introduced himself to me in front of my poster, I literally grabbed him by the lapels, dragged this giant of a man into a nearby room, sat him down, and announced: "You've got to do your film on the dolphin drive. And I'm going to tell you why."

❦

Taiji is a small coastal village that clings to the southeastern lip of the Kii Peninsula of Japan's Honshu Island, some 260 miles southwest of Tokyo. To the proud people of Taiji, population around three thousand, the broad Pacific Ocean has been for generations not only the source of their livelihood but also their sole medium of transport. In the past, the rugged, majestic terrain that's so typical of this part of Japan's main island made overland access to the coast all but impossible. Modern road and rail links now join this remote village to the rest of Japan and the rest of the world. Nevertheless, what goes on today in Taiji remained little known in 2005.

A source of pride of the people of Taiji lies in the town's history as the spiritual home of the country's whaling industry, going back to the 1600s. Indeed, visitors to the town see signs of that link everywhere, from the prominent whaling museum (built in 1969) to the images of whales and dolphins on billboards, buildings, and sidewalks. For centuries, with great ritual and considerable skill, the men of Taiji put to sea in swift, elaborately decorated boats to hunt whales, primarily right whales, which provided more than enough for the villagers' subsistence needs and plenty for trade to other parts of the country.

This long-lived tradition would not last forever. Their vigorous whaling activities and the incursion of other whaling nations into Japanese waters in the nineteenth century led to a decline in right whale populations. And a terrible storm in the winter of 1878 resulted in the loss of most of Taiji's whaling fleet and the death of its men, more than a hundred. That incident contributed to the breaking of Taiji's spirit as a whaling community.

A global ban on commercial hunting of large whales in 1986 formally ended the tradition for good. The International Whaling Commission (IWC) was created in 1946 under the International Convention for the Regulation of Whaling to provide for the proper conservation and management of whale stocks. Japan, Norway, and Iceland continue to ignore pressure to end commercial whaling and together kill over two thousand whales each year. But at least the IWC has a ban on the killing of large whales; there is no such ban on killing small cetaceans — dolphins and small whales. Herein lies Taiji's more recent, and less honorable, reputation. In Western eyes, Taiji has come to symbolize human disregard for other living creatures. The government of Japan allows the Taiji fishermen to kill twenty-three hundred dolphins and small whales each year in the most brutal and inhumane manner imaginable.

Every year, beginning in October and running through April,

a small group of fishermen, about thirty-four, sets out to sea from Taiji in a dozen or so motorboats to locate groups of dolphins. The fishermen position themselves between the groups and the open ocean and then herd the dolphins toward shallow lagoons using the *oikomi* method that we used to save Humphrey the humpback whale: The fishermen hold long metal pipes tipped with flanges over the side of their boats and rhythmically beat them with hammers. Underwater, the sound of the hammering is greatly amplified and travels far, a wall of sound from which the dolphins flee in terror. Frenetic and exhausted, they become trapped in a lagoon that the fishermen seal with a net across the entrance. Each drive hunt, as it is called, corrals social groups of dolphins, usually from twenty-five to two hundred individuals, including mothers and calves.

The dolphins are left in the cove, sometimes for days, until the next affront occurs. Trainers arrive from aquariums (excluding those in the United States, Europe, and Australasia) to inspect the milling dolphins and pick what they consider to be the crème de la crème for their dolphin shows. Flipper delighted children with its escapades on television in the 1960s, and since then, people's interest in watching dolphins' amazing physical skills and in swimming with them has grown. Aquariums, marine parks, and swim-with-dolphins programs make up a multibillion-dollar industry today. Here, then, is the motivation for the fishermen of Taiji: A dolphin sold locally as meat or for fertilizer will fetch a few thousand dollars, but a potential Flipper can haul in twenty-five thousand dollars or more. These "lucky ones" are shipped to destinations around the globe (Japan, China, South Korea, and the United Arab Emirates), forever separated from their natal social groups and their mothers.

The fishermen then herd the rest of the social group, the unlucky ones, to the killing cove in Hatagiri Bay, with craggy, scrub-covered cliffs rising steeply on three sides. Within minutes, the emerald green waters of the cove turn a deep crimson as blood gushes from spear wounds and deep gashes inflicted by long knives and harpoon blades. The terrified animals thrash and writhe in their own blood. Some try to escape, frantically swimming toward the head of the cove, losing blood as they try to breach the barrier, only to fall back, usually twitching a few more pathetic times, and then sinking below the surface. Many drown in nets that prevent them from reaching the surface.

The fishermen drag the wounded animals toward boats, using grappling hooks that often pierce eye sockets or breathing holes, tearing their living flesh. Years ago, in similar hunts in Futo, Japan, cranes hoisted animals by their flukes, leading to a kind of modern-day version of medieval torture. In water, dolphins are virtually weightless, and their anatomy is adapted to that state of near equilibrium. In air, gravity grabs at each one's eight hundred pounds. The dolphin is suspended by its tail, and its spinal column is stretched and ripped apart, causing pain we can only begin to imagine. (Something like this was done to traitors in the Tower of London in ages past, on a feared contraption called the rack.) Smaller knives are used next, as the fishermen eviscerate the dolphins, many of which are still alive. All the while the air is filled with the sounds of the fishermen shouting and laughing, of dolphin tails thrashing in water, of bodies flip-flopping on land in a dance of death, and a cacophony of dolphin vocalization. And all the while, too, the dolphins' "smiles" remain frozen on their faces, macabre masks of death.

The scene I just described has been witnessed firsthand by very few people not intimately connected with the Taiji fishing community. Townspeople go to great lengths to prevent prying eyes from seeing the slaughter. DANGER and KEEP OUT signs are posted at the few access points to the killing cove. Razor wire blocks off the unauthorized. Vigilant policemen turn back inquisitive foreigners carrying cameras. Large tarpaulins are hung as curtains across the cove to further restrict the view. But by the time of the 2005 San Diego meeting, a few intrepid activists had managed by ingenious means to capture the killing on film so that the full horror was beginning to be known. This was the scene I described to the silver-haired stranger who approached me at the San Diego meeting. He told me his name was Louie Psihoyos and that he'd been a photographer for seventeen years at the illustrious *National Geographic* magazine but that he had no serious experience in making movies. I knew him by reputation: he was someone with a knack of capturing the essence of a complex situation with a single, profound image.

Louie looked at my academically framed poster on distress calls and the drive hunt, and he seemed puzzled, not knowing quite what to make of it. I wasn't surprised; it *was* rather abstruse. But when I told him about the drive and showed him a short clip, he seemed to get it completely.

I explained to Louie that the townspeople of Taiji had recently become extremely defensive about the drive hunts. A growing international protest in the media and several attempts by activists to film or disrupt the drives had led to that. Doing a full documentary, I said, would present unusual challenges.

Taiji represented a very complex situation, and in the four years I had been working to stop the drives, my colleagues and I had suffered some setbacks. I had visited Japan two months prior to the marine mammal conference, having been invited to a small conference on vocal learning at Keio University in Tokyo to speak about vocal imitation in dolphins. I took advantage of this trip to meet with the director of the Nagoya Aquarium, one of the largest and most modern aquariums in Japan, and give a talk to its marine mammal staff. It was hard for me to believe that this aquarium procured its dolphins from the drive hunts, especially since its director was a well-known turtle conservationist. Paul Boyle and I thought that perhaps such a meeting would serve to open communication about the dolphin drives and that we might be able to find a Japanese ally, a conservation-minded leader in the aquarium community who would spearhead local efforts to end the hunts. I spoke privately to the director and asked him to explain the rationale for continuing the drives. He turned to me, seemingly uncomfortable, and said that the aquarium was subsidized in part by the govern-ment of Japan and that it had instructed the aquarium to take dol-phins from the drive. The government considered dolphins com-petitors for fish, and so they had to be exterminated. The director said of the drive hunts, "It is a pest-control operation, because the dolphins are depleting local fish stocks." There is in fact no scientific support for this claim, and overfishing is much more likely to be the cause of declining fish populations.

On the national stage, various authorities offered different ra-tionales. In response to the assertion that the animals suffer pro-longed pain, Jun Koda, a counselor of the Japanese embassy in Lon-don, said that the fishermen were careful to minimize pain, and every dolphin "almost instantly meets its end within a maximum of 30 seconds and does not suffer any pain."[1] As I told Louie, this was obviously not true, as even the shortest of video clips attested. The dolphins were eviscerated alive, suffering very slow deaths.

Cultural differences are often raised to deflect criticism. "If someone eats a cow, why should one object to a dolphin being eaten?" asks Hideki Moronuki, an official in the Ministry of Ag-riculture. "They're all mammals."[2] This is true, now many slaugh-terhouses make efforts to minimize pain and suffering; there are growing efforts to minimize pain and suffering in farm animals as well. Many improvements are needed, but dolphins are within my area of expertise, and their welfare is my concern. I want to end the drive hunts, period. These animals need global protection. But the manner by which an animal is killed is certainly relevant in the overall argument. The government of Japan's final defense of the hunts amounts to "It's part of our culture." "The feeling here is that the world needs to respect cultural differences," Tetsu Sato, a pro-fessor of environmental management at Nagano University, told a reporter for the New York Times in the fall of 2009. "Why should there even be a debate on this issue?"[3] There is a debate on the issue because the issue is not about cultural differences. It is about the inhumane treatment of intelligent, sentient creatures. We don't accept cultural differences in the mistreatment of humans, and it is no better an excuse for mistreatment of sentient animals.

Louie listened patiently and intently to my case, and I showed him videotapes of the drive hunts. I then introduced him to Hardy Jones, an environmental filmmaker who'd devoted much of his later career to the welfare of dolphins and whales. That evening we attended a short film of Hardy's on the drive hunts. It had been Hardy who, four years earlier, had effectively recruited me to join the campaign to end the Taiji drive hunts. When Louie and I parted, he told me that he was interested in the prospect but couldn't say for sure if he would be able to do it. He would let me know.

◈

A year and a half before the San Diego conference, I had faced my greatest setback yet in the campaign. The rain was sheeting down as Paul Boyle, Steve Olson — the vice president of government affairs for the Association of Zoos and Aquariums — and I were chauffeured north along the ultra-chic Embassy Row in Washington, D.C. We were nervous, or at least I was, but we had a degree of optimism despite the rain, an expectation that carefully presented scientific argument would generate genuine insight and lead to a collaborative path toward the success of our cause. It was the last day of March in 2004, Steve's birthday, as it happened, and as our driver stopped at 2520 Massachusetts Avenue, the embassy of Japan, we looked at one another and silently agreed: *Okay, this is our one big chance!*

Our goal was straightforward. We aimed to persuade the embassy's first secretary of fisheries that the annual drive hunts of dolphins in Taiji were patently inhumane and should be stopped and then have him transmit this message to the office of the prime minister in Tokyo. When I'd arranged the meeting I was told to bring any scientific evidence I had to make my case.

Given my theatrical background, I planned to use the *Miracle on 34th Street* model to achieve the goal. In the 1947 film by that name, Kris Kringle, a kindly old man who believes himself to be Santa Claus and plays that role in the Macy's department store in Manhattan, is committed to an institution because of his insane belief. Judge Harper demands that Kringle's lawyer produce material proof that Kringle is indeed who he claims to be or the old man will languish in the institution for a very long time. A little twist of Hollywood storytelling allows the lawyer to place before the judge fifty thousand letters addressed to Santa Claus, all delivered to Kringle by the U.S. Post Office. Faced with such massive evidence, the judge feels compelled to rule that Mr. Kringle must indeed be who he says he is.

Miracle on 34th Street was fiction, whereas the drive hunts at Taiji were unrelenting reality. And the letters to Santa proved a fiction within the fiction, whereas we aimed to prove scientific facts. But I liked the idea that overwhelming evidence had the power to convince even the most skeptical judge. And the Japanese authorities had shown themselves to be skeptical about protests of the Taiji drive hunts. Our massive evidence that day at the Japanese embassy was contained in three very large boxes of scientific papers labeled, respectively, ANATOMY AND PHYSIOLOGY, SOCIAL BEHAVIOR, and INTELLIGENCE. Collectively, these three boxes of learned articles represented everything known to science about dolphins.

We were convinced that this body of evidence showed dolphins to be creatures with intellectual, social, and emotional lives on at least the level of chimpanzees, and in some cases even more highly developed. If it was true, and it was, then the drive hunts at Taiji were grossly inhumane. We hoped that, like Judge Harper, the first secretary would be convinced. After all, Japanese primatologists were leaders in research on chimpanzee intelligence and behavior, and great apes were revered in Japanese culture. The slaughtering of apes was illegal, and only poachers in Africa attempted it. It seemed to us that dolphins were not revered in the same way as apes because people lacked the relevant information, and we were going to provide that information.

We were met by a lawyer, Gavin Carter, who represented Japanese interests at meetings of the International Whaling Commission. He was very British and had a kind face, but he was like a rock underneath. I had come to this meeting at the embassy with good advice from my uncle, who had been the head of international trade with the East during the Eisenhower administration. "Negotiations with the Japanese are different than with Americans," he'd told me days earlier. "In the States, you don't have to know someone in order to close a business deal. Not so in Japan. But once you have a relationship with someone, then you can do business." We didn't have much time to get to know the other party, but I was determined to reach out and establish a rapport, not treat the meeting as a confrontation. We were determined to avoid being the ugly Americans, dictating to other countries how they should behave. We wanted to share our science, and then we hoped the government would want to do the right thing.

The first secretary entered, immaculately turned out, like everyone and everything around us. For so senior an official, he was young, much younger than I had expected, perhaps in his thirties. And at no more than five foot seven, this young man didn't tower over me the way most of the men I encountered did. He extended his arm, we shook hands; he bowed slightly in the formal Japanese way, and I lowered my head too. To my surprise, I saw that this senior bureaucrat was wearing black sneakers. This incongruity caused an irrational thought to pop into my head: *Maybe we have a chance!*

❧

The first secretary led Paul, Steve, and me to a large conference room dominated by a vast rectangular table. The room was windowless and quite dark. As we had requested, it was set up for a

PowerPoint presentation, with a screen at the far end of the table. We put our *Miracle on 34th Street*–style evidence, the three boxes of academic papers, on a side table. I hooked up my laptop to the projection system and prepared to open the meeting with a rapid excursion through what we knew about dolphin intelligence, communication, and social behavior.

We were actually a little surprised to find ourselves in the embassy that dreary March day because the notion had only entered our thoughts a bare few months earlier. I had been giving a paper at another MMS conference in December 2003. Right after the paper's presentation, Naomi Rose and Courtney Vail came up to me to discuss the drive hunts.

Naomi was a senior scientist for the Humane Society International, specializing in global marine mammal protection issues, and Courtney was the U.S. policy officer for the British-based Whale and Dolphin Conservation Society. They had a joint booth at the conference, an indication of how the dolphin killing in Taiji was bringing together organizations and people from different backgrounds in a unified effort that was truly remarkable. Naomi and Courtney told me that they had tried to get a meeting at the Japanese embassy in Washington, D.C., to press their case. "They would only agree to meet us at five o'clock," explained Naomi. "The embassy was just closing for public business for the day, so we knew it was a brush-off. They wouldn't even let us through the front doors!" Courtney told me that she and Naomi had wanted to give the embassy official just one scientific paper that spoke to dolphin intelligence and the issue of the inhumanity of the drive hunts. "We thought that your paper with Lori on mirror self-recognition was the most powerful one," Courtney said. "We had it with us. They took it and told us that while they were not prepared to talk to animal rights groups or environmental groups, they would be prepared to meet with the author of the paper, as a scientist. You just need to call them to set up a meeting."

Our principal concern in planning for the meeting was staying on message: the drive hunts at Taiji were brutally inhumane and should therefore be stopped. Paul was especially insistent that we had to, as he kept putting it, "stay on this square" during our projected meeting. Straying into any other area — cultural traditions, sustainability of the hunts, whether dolphins should be killed at all — immediately allowed the Japanese to throw up a smoke screen that obscured the main issue. We recruited Steve Olson, with his expertise in government affairs, into our discussions, and eventually to our delegation, because he brought bureaucratic savvy and gravitas to our mission.

I titled my presentation "A Mind in the Water: A Review of the Scientific Literature on Dolphins." It was a celebration of dolphin intelligence in many realms — their brains, their awareness of self, their sentience, their social complexity. I stressed that even though dolphin and primate bodies were very different from each other in form and evolutionary history, dolphin and primate minds were very similar. I felt it was critical to present a portrait of the dolphin in a new frame. I acknowledged that the Japanese scientists were leaders in the field of ape cognition and pointed out that dolphins were cognitive cousins of the apes. It was quite an excursion in forty minutes.

I stuck almost exclusively to the science, but I did show some footage of drive hunts. That was a very special moment, because I knew that no one else had been able to do that in such a context. I also played them the plaintive distress calls of individual dolphins during a drive. And I showed a two-page spread of the blood-red waters of the cove at the island of Iki that had been published in *National Geographic* magazine in 1979. I finished the presentation with the following statement:

> We respectfully ask the Japanese government to take action in stopping the inhumane practices of the dolphin drive fisheries in Japan. Scientists have now found many parallels in the social complexity and cognitive abilities between the great apes and

dolphins. Recently, dolphins have been shown to possess a high level of awareness previously observed only in humans and the great apes. While great apes are revered and appreciated in Japan, dolphins — cognitive cousins of the great apes — are still harvested as a resource. Please let science guide your future policy regarding drive fisheries in Japan. We ask that you lead the way and take action in stopping this inhumane practice immediately.

As I'd been giving the presentation, I'd tried to get a sense of my audience. Were they really listening? For the most part, it was hard to discern what either one of them was thinking. The footage of the drive hunts did elicit some involuntary giggling, a typical Japanese response to embarrassment, as I later learned. Gavin Carter's face began to redden, but I didn't know what to make of it.

When I finished, I politely thanked the first secretary for listening. Paul began to draw the link between the science that had just been presented and the inevitable conclusion that drive hunts were brutal and inhumane, just as we had planned. But he didn't get very far. "Dr. Boyle, how can you tell us to stop this practice at Taiji when the United States is a whaling nation?" interrupted the first secretary, his jaw tight. These were the first words out of his mouth. We were caught completely off-guard. "What are you talking about?" Steve asked. "Your own Native Americans kill whales in the Northwest," the first secretary shot back. He was apparently talking about the Makah Indians. The issue of whether the Makah Indians should be allowed to hunt gray whales had been a contentious one for many years, politically and legally.

An 1855 treaty with the United States allowed the Makah to hunt whales as part of an agreement to cede their rights to large segments of their land. But late-twentieth-century global sensitivity to killing whales had prompted opposition to the treaty, and a legal battle had ensued. The Makah eventually won their case (with their legal fees provided by the Japanese government). Steve responded to the first secretary by pointing out that the Makah hunted for subsistence only, that very few animals were killed each year (it was limited to five under a new agreement), and that attempts were made to minimize suffering (the whale was harpooned by traditional methods and then immediately killed with a high-powered rifle). Here we were, not five minutes into the postpresentation discussion, and the Japanese official had already pulled us off Paul's square.

There followed a little more than two hours of a very polite, but increasingly tense, tug of war. Every attempt on our part to return the discussion to the issue of the hunt was met on their part by detailed and often unanswerable questions. They asked which animals should be eaten and which should not be; they asked about humane killing practices. They brought up issues about the Makah. We were in danger of losing the focus of our meeting.

We also discussed our belief that neither aquariums nor dolphin swim programs should be procuring dolphins from the hunt, and in fact, that very day, the board of directors of the Association of Zoos and Aquariums issued the following resolution: "Zoos and aquariums accredited by the Association of Zoos and Aquariums (AZA) are experts in animal care, wildlife conservation and educating the public about wildlife issues. The AZA strongly believes that the killing of dolphins and whales in drive fisheries is inhumane and should be terminated immediately."

Afterward, Paul described the Japanese tactics as the black-smoke effect. "It's like you can't see across the room, because of all this black smoke that has been deliberately created." It was all very respectful and businesslike. When it was over, Steve politely thanked the first secretary for agreeing to speak with us. The first secretary bowed slightly and said it had been a pleasure. We asked for a timely response from the Japanese government and agreed to supply more information to the embassy if requested to, and the first secretary agreed to draw up a list of further questions for us.

In hindsight, we had been naive in the extreme, practically delusional. We had expected a moment of enlightenment to arise from this one small face-to-face meeting. What had we been thinking? And why had they even bothered to meet with us at all?

There was a palpable silence from the embassy for the next three months. Finally, after Steve called the embassy, Gavin Carter e-mailed us with the promised list of questions. As had happened during the discussion at the embassy, the questions were so convoluted and vague as to defy a ready response. How intelligent did an animal have to be before it was no longer ethical to kill it and eat it? Why was it all right for Native Americans to kill a small number of whales? Did the number of animals killed affect how ethical a particular hunt was? Did we plan to stop other nations from killing whales too? Steve described the collection of questions as being "masterfully done, with no simple answers to anything." Paul and I were willing to respond to the letter, and in fact drafted some answers. Ultimately, Steve persuaded us that it would be a waste of our time. Just as the meeting at the embassy had done, Gavin's letter sidestepped the single, simple question at the heart of the Taiji drive hunts: Is the *oikomi* drive and subsequent killing inhumane? No amount of intellectual contortion on our part over questions that were essentially irrelevant would get them to focus on that point. We collectively decided not to respond.

Over the following eighteen months, the World Association of Zoos and Aquariums joined the Association of Zoos and Aquariums in issuing a formal resolution condemning the drive hunts at Taiji. The Act for Dolphins website that Paul Boyle, Lori Marino, and I established gathered two hundred and fifty thousand signatures on a petition to stop the drives, and it garnered some coverage in the media. A press conference that we orchestrated through Act for Dolphins was held at the National Press Club in Washington, D.C. in 2005; all the most important scientific voices against the Taiji drives were gathered, but the event was all but eclipsed by the timing of a major veto on stem-cell research by the Bush administration.

Finally, a couple of months after the 2005 Society for Marine

Mammalogy meeting in San Diego and almost exactly one year after our ill-fated visit to the Japanese embassy, a glimmer of hope broke through the dark clouds. "Hello, Diana, this is Louie Psihoyos," said the voice on the phone. I was standing in the foyer of our apartment on Riverside Drive in Manhattan. Several months had passed since I had pleaded my case to Louie in San Diego, and I had almost given up hope that anything would come of it. "Diana, I'm in New York," Louie continued. "I'm at JFK airport. I'm on my way to Japan. We're going to do the film. Wish me luck!" For the first time in months, I felt truly exhilarated and once again hopeful. I punched the air in the way young people sometimes do when they're really excited and shouted, "Yes!" Juvenile, I know, but that's how I felt.

The project on which Louie embarked that day was no ordinary documentary, for several reasons. First, with Jim Clark's fortune* bankrolling the effort, the final product would be as slick as any Hollywood feature film, complete with special effects. Second, with the fishermen, the town council, and the town's police force on high alert to prevent access to the killing area by whatever means was necessary, Louie's team found themselves under more than the usual degree of production stress. "Four years ago, through Jim, I met Steven Spielberg," Louie told an interviewer for *indieWIRE* at the beginning of 2009. Louie went on to say that, thanks to his experience with *Jaws,* the filmmaker had advised him never to make a movie that required boats or animals, and Louie now had his own advice to add to Spielberg's: "Never make a movie where the subjects want to kill you and you have to work in the middle of the night to break the law while the police are on your tail." That had been Louie's experience in making *The Cove.*

---

* Jim Clark, the founder of Netscape, was a diving partner and friend of Louie, and together they founded the Oceanic Preservation Society, in 2005. It was Clark to whom Louie referred when he said he had access to someone with as much money as any shah.

The fishermen's determination to prevent Louie from getting access to the killing cove was just one obstacle. A second was its fortresslike topography: steep cliffs on three sides and open ocean on the fourth. A team of Navy SEALs would have been helpful, but the production team didn't have that option. So Louie turned to his longtime diving friends Mandy-Rae Cruickshank and her husband, Kirk Krack, to swim into the cove, plant underwater cameras and microphones, and then return undetected, all without benefit of scuba equipment. Cruickshank and Krack are no ordinary divers. They are among the world's best free divers; that is, they dive to great depths without using any breathing apparatus. Cruickshank is capable of diving three hundred feet and is able to hold her breath for as long as six minutes, a near-dolphinlike performance!

Louie persuaded his first photo assistant—the head mold maker at Industrial Light and Magic, George Lucas's legendary special effects company—to make fake rocks with high-definition cameras and microphones embedded in them. Intrepid team members, wearing camouflage and military-style face paint to elude the ever-vigilant police, installed them on the craggy cliff sides near the head of the cove. (Louie said that the fake rocks were so good that his team had a hard time finding them after the filming was done.) Another critical team member was Simon Hutchins, a former Canadian Air Force avionics technician; he organized the underwater expeditions and was the technical wizard behind the drone-carried, gyro-stabilized camera. A team of expert surfers, led by the legendary Dave "Rasta" Rastovich, and a clandestine-operations organizer completed the production company. A motley crew, you might say. "We're all professionals," observed Hutchins, "just not at filmmaking." Louie called his crew the Ocean's Eleven, a reference to the 1960 Rat Pack caper film and the 2001 remake of that name.

The Cove was released in the fall of 2009 to instant acclaim. There was some criticism that it was "too evangelical," which was not surprising because it had been Louie's stated aim to "put a stop to the most ghastly slaughter of animals on the planet." It garnered a top prize at the Sundance Film Festival and the following March won the Academy Award for Best Documentary Film at the 2010 Oscars. In the ninety-minute film, there were fewer than two minutes of footage showing the actual killings; it was what one might call a Disney version of what really went on every year at Taiji. Louie noted that he had much more horrifying footage, but it was simply too graphic to show. As gruesome as it is onscreen, the reality is worse, I can assure you. I was a science adviser for the film and was delighted that it did so well.

The only problem I had with the film was that it carried more of an anti-aquarium sentiment than I thought it should. I was trying hard to build the strongest possible international movement to bring an end to the drives. My goal was (and still is) to unite the largest possible consortium of scientists, animal-care professionals, and animal-welfare NGOs such as the Humane Society of the United States, the Whale and Dolphin Conservation Society of the UK, and the Earth Island Institute in San Francisco in a science-based call to stop the slaughter. The taking of dolphins from the drive by any aquarium or organization for commercial, educational, or research purposes is unconscionable. Yet the film attacks *all* aquariums in one swooping condemnation, suggesting that all aquariums obtain their dolphins from the drive. This is far from the truth. Just to set the record straight, aquariums in the United States are committed to caring for the dolphins in their facilities. They participate in cooperative breeding and artificial insemination programs, moving sperm rather than animals to maintain a healthy genetic pool. I am unaware of any U.S. aquarium that has procured a dolphin from the wild in more than twenty years.

The oversimplifications and implications in the film alienated many aquariums, which are important allies in the mission to end the drives. The film, ironically, damaged the overall cause — and weakened, at least temporarily, our coalition.

Changes in thinking, cultural and scientific, often take time. We need a sea change in how dolphins are viewed in the parts of the world that continue to regard them as commodities, as expendable, and even as pests to be exterminated. Most aquariums in the United States (and in a few other countries) place high value on healthy dolphins living in a social network of mothers and young and other pool mates. These aquariums need to increase their pressure to stop those renegade aquariums that still take dolphins from the wild.

The Cove was shown at the Tokyo Film Festival in October 2009 to a muted response. Plans for wider release in the country met with vigorous protests from nationalist groups declaring that any movie-house owner who showed the film was a traitor to Japanese honor. Violence was threatened. Eventually, half a dozen movie houses showed the film; no violence occurred. Nor was there a popular uprising against the killing — or at least, not as of the closing weeks of 2010. The Taiji Municipal Council arranged a carefully controlled, media-free meeting with a selected group of animal rights activists that November. The townspeople stuck to their position. Although the drive was suspended for a few days in the September after the film's release, it has been business as usual ever since. The council's president said at the November 2010 meeting, "We believe that these are natural resources, to be used effectively." He still doesn't get it! But I, for one, am not giving up.

10

ENDING THE LONG LONELINESS

> One does not meet oneself until one catches the reflection from an eye other than human.
>
> —LOREN EISELEY

I HAVE TAUGHT a course for many years now, variously called *Animal Communication and Cognition, Animal Minds,* and *Communication between Humans and Other Animals.* No matter the title, the purpose is always the same: to explore how we look for intelligence in other animals and ask an old question: Do animals think? I start the course talking about how as early as the fourth century B.C.E., Aristotle had classified the natural world and described man's relationship to other animals in his concept of the Ladder of Nature,* a Platonic version of souls in which plants were endowed with vegetative souls, responsible for reproduction and growth; animals (a class that did not include humans) had both vegetative and sensitive souls, which oversaw mobility and sensation; while humans, in splendid isolation, were the sole possessors of rational souls, which endowed us alone in nature with the capacity for thought and reflection.

We journey onward into the thirteenth-century world of Saint Thomas Aquinas, who accepted the Aristotelian scale of being, yet, foreshadowing Darwin, also envisioned a degree of continuity between humans and other animals. Although he saw man as alone

---

* This was the basis of the Great Chain of Being, a conception of nature that went from the inanimate to the sublime, with man being just a little lower than the angels; it dominated Western thinking for centuries and its echoes linger still.

and superior to animals in having an intellective soul, he saw the powers of man as not so very different from those of other animals, only more "heightened." Aquinas's dualistic view of men and animals, all of whom combined physical bodies with ethereal souls, had a major impact on the writings of René Descartes in the seventeenth century, who denied such dualism for other animals. He stripped the soul from nonhuman animals and left it in the sole possession of us. Only humans had a thinking substance — a ghost in the machine. All other animals were automatons merely sleepwalking through life, aware of nothing, thinking nothing. This view dominated until the nineteenth century, when Charles Darwin and Georges Romanes enthusiastically embraced the belief that nonhuman animals were indeed capable of both rational thought and emotional life, even if not as lofty as our own. Yet this aspect of Darwin's thinking was harshly dismissed by the school of behaviorism beginning in the 1920s, which effectively catapulted nonhuman animals back to Aristotle's exile from the club of rationality. At best, the question of animals' minds was considered beyond the pale of science because, behaviorists said, thoughts and feelings in animals were private phenomena and therefore inaccessible to objective measurement, so it was foolish to waste time trying to access them.

The cognitive revolution of recent decades, pioneered largely by Donald Griffin, has brought forward a new perspective on animal minds, one with which Darwin and Romanes would have been comfortable. We now recognize that *Homo sapiens* ("wise man" or "knowing man" or "thinking man") shares the world with other creatures that think too. By the end of the course, my students come to see that the initial question, *Do* animals think?, is the wrong question. It should be, *How* do animals think? Animals are capable of far greater richness of behaviors than was once imagined; they were simply *assumed* to be incapable of thinking, to lack minds of a kind that resembled ours in *any way*. If you live with a dog or a cat or if you ride horses, you may be quick to say, "I already knew they could think!" But the path that I trace with my students requires rigorous science rather than affectionate anecdotes. And at the end of my course, I leave science behind and turn to poetry.

Two fine thinkers have had important influences on my work, and I feel a visceral resonance with their ideas. One is the British anthropologist and social scientist Gregory Bateson, whose phrase *pattern which connects* speaks eloquently of our connectedness with nature and other species. The second is Loren Eiseley, American anthropologist, philosopher, and natural science writer, whose prose matches or even exceeds the luminosity of his thoughts. It is to Eiseley that I turn in the final class of the course, first with his poem "Magic."[1] Eiseley received thirty-six honorary degrees during his varied career, making him the most honored member of the University of Pennsylvania since Benjamin Franklin. He was devoted to bringing science to life for the general public.

In "Magic," Eiseley described how he became enchanted by a particularly vibrant male red cardinal, one that liked "practicing vocal magic" as he flew back and forth by the windows of Eiseley's house. Eiseley described the morning ritual of opening the kitchen window, placing seed on the sill, closing the window, and waiting for that red male, along with his family, to come to feed. Tentatively at first, the birds soon came to demand the morning offering, expressing impatience if Eiseley was late or forgot, which he rarely did. Eiseley felt affection for the entire family, but it was that first red bird that found its way into his soul as the two individuals — bird and human — forged a special relationship. He thought of the bird as a "sorcerer," and he was its "apprentice." This was not the role of the brash young sorcerer's apprentice in Goethe's poem, written in 1797, who, tired of the mundane chore of cleaning his master's workshop, tried to enlist his yet-to-be-tamed magical powers, to disastrous effect. Instead, in Eiseley's eyes, being a sorcerer's apprentice meant he was in a position to learn some of the magic the sorcerer wielded. Alas, before many weeks passed, the sorcerer apparently met with an accident, and its nest was thereafter deserted.

Two lines in the poem have special meaning for me, as they speak to my own experience:

I love forms beyond my own
and regret the borders between us.

I, too, have developed rituals of the sort Eiseley describes in "Magic," feeding the cardinals and imperious blue jays in my garden in Connecticut. But it is the antics of the crows that draw me. If I hadn't found my way to studying dolphins, I think I would now be working with crows. "I love forms beyond my own": Ever since I was a young girl I've felt an extraordinarily urgent, and to me entirely natural, connection with animals. My pets, and the waifs and strays that I was constantly rescuing, elicited compassion in me as instinctive as the urge to climb trees and other tomboyish antics of childhood. And I really felt I could communicate with my dog and with other creatures, as if I could wield the magic of King Solomon's ring. Show me a kid who, given the right environment, doesn't feel that way, who doesn't know in her being that she and the creatures of the world of nature are one. It is, in the purest and most innocent sense, the experience of a strong connection, a pattern that connects and a deep sense of caring — I can only describe it as unconditional love. Too bad that as we grow older and become immersed in the trappings of civilization, most of us are oblivious to its gradual disappearance.

The German-born American social scientist and philosopher Erich Fromm coined a word for this powerful connection with nature: *biophilia,* a love of life and living systems. In a book with that single word as its title, the evolutionary biologist and naturalist E. O. Wilson argued that biophilia is encoded in our genes, the product of evolutionary interdependence in our Paleolithic ancestors. "[W]e are human in good part because of the particular way we affiliate with other organisms," he wrote. "They are the matrix in which the human mind originated and is permanently rooted, and they offer the challenge and freedom innately sought. To the extent that each person can feel like a naturalist, the old excitement of the untrammeled world will be regained. I offer this as a formula of reenchantment to invigorate poetry and myth: mysterious and little known organisms live within walking distance of where you sit."[2]

For many of us in technologically advanced societies, including many environmentalists, the reductionist ethos of science that breaks nature into its component parts and doesn't see the whole, coupled with our innate natural bias sometimes called speciesism, leads us to view nonhuman creatures as inferior to *Homo sapiens,* beings to be judged by their economic usefulness to us, with no intrinsic value of their own. The reality of interdependence among all of Earth's organisms has little place among the essentially mechanistic mainstream thinking of today.

For most of us, then, biophilia, part of the fabric of what it is to be human, is glimpsed only occasionally, when you stop in your tracks to gaze at a glorious natural panorama, or when you take time to walk in the woods, looking, listening, smelling nature. Or when you gaze into your pet's eyes and recognize a returned gaze. I feel extraordinarily privileged to have been able to enter the world of dolphins so intimately over the years, in my work with them in aquariums, during rescues, and in their natural habitat, which reinforces my sense of a unity in biology through the patterns that connect that I see every day. It keeps me connected to that kid in me. And in this I resonate with another line in Eiseley's "Magic":

How does a man say to his fellows

        he has been enchanted

                by a bird?

And how does a woman say to her colleagues, and to the world, that she has been enchanted by a dolphin? Just as Eiseley allowed himself to be taught the magic of his little avian sorcerer, not only have I been enchanted by Circe and her successors, I have been taught by them, taught to see their world through different eyes. In his book *King Solomon's Ring,* Konrad Lorenz spoke of the supposed

magic of the seal ring as a metaphor for enhanced powers of observation of animals' behaviors and modes of communication. Those
of us who study animal communication are effectively in search
of King Solomon's ring, in search of keener ways of observing and
understanding them. That is what Circe and her successors have
taught me.

⌘

Consider the stories of dolphins supporting injured or aging fellow dolphins, preventing them from sinking and drowning, or approaching sailors and others in trouble at sea, warding off sharks or
guiding distressed people to shore and to safety. If these acts were
carried out by people toward other people, or toward animals in
need of help, we would describe them as showing empathy and
compassion, an expression of felt care for another individual. The
question is, when dolphins behave in these ways, do these actions
also come from empathy, compassion, and care? In other words, do
they know what they are doing, or are they mindlessly following a
primal drive?

These are difficult questions; it is hard to know what is in the
mind of an individual with whom you are not able to converse. But
as Circe showed with Delphi, not every mother has an unerring instinct or understanding to push her poor flailing calf to the surface.
And Circe wasn't the only mother I saw who apparently had no
push-the-flailing-baby-to-the-surface instinct. There are enough
such observations that I believe the primal-drive argument fails.
Which leaves us with a different question: How do dolphins think
under these circumstances? We have ventured along this path before, of course. But here I want to use the question to explore some
of the resistance to the idea of animal thinking.

In 1960 Eiseley wrote a beautiful essay called "The Long Loneliness,"[3] which begins "There is nothing more alone in the universe
than man." Our supposed "loneliness" was the result of the wide-

spread belief of the time that only humans possessed rational, thinking minds and high-caliber intelligence. This left our species alone on an intellectual pinnacle, unable to communicate our thoughts and feelings to any other creature despite a strong desire to do so. "When we were children," he wrote, "we wanted to talk to animals and struggled to understand why this was impossible." (Some of us kids, of course, believed we could!) "Slowly we gave up the attempt as we grew into the solitary world of human adulthood; the rabbit was left on the lawn, the dog was relegated to his kennel."[4]

Eiseley's inspiration for the essay had been John Lilly's work on dolphins and his ideas about their large, complex brains and inferred high intelligence, which offered a possible pathway to ending the long loneliness. Eiseley argued that perhaps we hadn't recognized other intelligent minds here on Earth because we had been using a human model of intelligence — symbolic language, hands capable of making tools, and so on. What if we imagined an intelligence such as ours in a creature that had exchanged hands for fins and lived in the ocean? What would the evidence of intelligence look like in the absence of the products of science and technology, in the absence of material manifestations of our minds?

I read the essay in my early days as a graduate student, and it was very influential in my thinking. Yet I have never experienced the kind of loneliness to which Eiseley refers. I have always felt an abundance of life and intelligence around me in nature. But he asks the key question: What kind of experiments do you do if you don't know what you are looking for? Donald Griffin's discovery that bats navigate using echolocation is a good lesson. At the time, no one had even considered that animals might use a form of radar to find their way in the world, and so no one had actively looked for it. Griffin stumbled on it by chance and was greeted with disbelief when he reported what he had found.

When we try to interpret animal minds, animal thinking, and animal behavior, it is all too easy to use our own experience as a model. This act of anthropomorphism, of assigning human qualities to animals, can be a helpful start to understanding what is going on in the mind of an animal, but it is also dangerously seductive. As a scientist I must find a balance between anthropomorphism and anthropocentrism, in which we assume that we humans alone are unique in our abilities and that our kind of intelligence is the only "real" intelligence. In the writings of animal behaviorists and animal intelligence researchers over the past eight decades or so, there are many stern warnings about the pitfalls of anthropomorphism. But there was little, if anything, about the dangers of anthropocentrism until the beginning of the cognitive revolution, precisely because it governed a great deal of mainstream thinking. Anthropocentrism remained an unspoken assumption of reality: that *Homo sapiens* were so different from other animals, so special, that no useful parallels between ourselves and any other species could be drawn.

One of my favorite snippets from history on this topic concerns the wife of the bishop of Worcester. In 1860, following a debate at Oxford between Thomas Huxley and the good bishop on the matter of Darwin's theory of evolution by natural selection, the bishop's wife purportedly said, "My dear, descended from the apes! Let us hope it is not true, but if it is, let us pray it will not become generally known." The notion of any connection between exalted humanity and the base world of animals was, and is, abhorrent to many people, then and now.

Evolutionary continuity of physical form between humans and anthropoid ancestors, and by extension the living great apes, has long been an accepted fact. Yet even in physical forms, anthropocentrism lingers on. The evolutionary tree of life is envisaged by some as a conical pine, with humans at the very top of the tree — the pinnacle. This is not so unlike Aristotle's placement of humans, alone, at the top of the Ladder of Life — alone. The reality, as all evolutionary scientists know, is that the tree of life is more like a gracious elm, broad and lush, with *Homo sapiens* one branch among many others, each of which has its own collection of ingenious adaptations. Man is no longer seen as the pinnacle of evolution but rather as one outcome of a diverse process of adaptation.

Yet despite the acceptance of physical continuity between humans and the rest of the animal world, the idea of cognitive continuity has been stubbornly resisted. As leading behaviorist Lloyd Morgan put it: "In no case is an animal's activity to be interpreted in terms of higher psychological processes, if it can be fairly interpreted in terms of processes which stand lower in the scale of psychological evolution and development." In other words, scientific logic supposedly demanded the most parsimonious conclusions about animal behavior — that it is the product of mindless, unconscious responses to stimuli in the environment rather than of conscious thought.

Parsimony is generally a good strategy, but it can be a problem if applied in the wrong way. Viewed in this manner, much of our own behavior could be interpreted as mindless responses to environmental stimuli rather than the product of active thought. And indeed, one can engage in a behavior while switching in and out of conscious thought about it — breathing, for example, or taking a walk. There is no clear method by which an observer could differentiate between a person's conscious and unconscious modes. For animal researchers, everything depends on what we are looking for, how we are looking, and what we expect to find. This brings us back to Eiseley's poem in which he regrets "the borders between us." What if the borders between us are not real but merely assumed, based on lack of evidence? Shouldn't we sometimes allow for the *possibility* that what a certain behavior, such as an act of apparent empathy, looks like is in fact what it seems?

Some animal-thinking deniers, such as the British Darwinian philosopher Helena Cronin, have argued that the great flexibility and versatility in animal behaviors that have been uncovered during the decades of the cognitive revolution are best explained as the outcomes of genetically determined biological programs. Just as computers can be programmed to carry out complex functions, she argued, so too can animals be programmed (in their DNA) to be behaviorally complex in the absence of conscious decisions or conscious experience. But, as Griffin pointed out from the beginning of his writing on the subject, an animal that has some degree of cognition will be more flexible and more efficient in the face of myriad environmental challenges. "This is the most parsimonious position," he said. "I'd say it is so plausible that I would put the burden of proof on Dr. Cronin, or anyone else, to tell me why it would not be useful for animals to think consciously."[5] I agree.

The issues of empathy, compassion, and care are even more contentious than, say, whether a young baboon makes a decision to try to sneak a quick copulation with a desired female while the alpha male is elsewhere. More contentious, because empathy and compassion are considered to exist at a higher level of cognition than decisions about foraging and mating. They seem much more human, if you will. What, then, can we say that might encourage us to believe that dolphins experience something akin to empathy when they are impelled to save a drowning sailor?

We know from the mirror self-recognition experiments that dolphins have a sense of self, which is one requirement for having a sense of what another is experiencing. But I would also point to the fact, and I use the word *fact* purposely, of cognitive continuity. Human brains and the brains of other animals are built from the same components, so there is at least the possibility that the functions they support will be similar. Griffin pointed this out in his first book, *The Question of Animal Awareness,* published in 1976, and has done so repeatedly in subsequent books and many papers. Here's a recent example: "In view of the similarity of neurons and synapses, and the flexibility of many animals' behavior, it seems unlikely that the difference between human and animal minds is an absolute dichotomy, with no animal ever conscious. Instead, the principal difference is probably the *content* of consciousness."[6]

What that difference is, we cannot know objectively. Unlike

Gordon Gallup's mirror test for identifying self-recognition, there is no equivalent, simple, objective test for empathy. I know what I *feel* when I am moved to an act of compassion, and I can guess what you feel under similar circumstances. Something related might be occurring in the minds of dolphins when they perform acts we would describe as functionally empathetic. And if so, it is possible that recipients of acts of compassion understand and appreciate it in some way too. To make this suggestion is not an emotional leap; it is a logical leap, based on a degree of cognitive continuity.

Recall that moment when Humphrey came back to our boat at the end of our rescue effort and stared up at us for several lingering seconds. There was definitely something on his mind that compelled him to do that.

∽

Conversely, imagine what might be going on in the mind of a dolphin as it is being eviscerated alive by a fisherman during the annual drive hunts at Taiji. Frankly, I think about the dolphin hunts every day, and I am determined to find a means to bring them to an end. I am passionate, even obsessed with this. How could I be indifferent? My science has brought me to the realization that dolphins are highly intelligent, feeling creatures that, like us, have a sense of self and a capacity for care and compassion. Killing these sentient animals as the Taiji fishermen do is brutal, inhumane, and unjustifiable.

When I first started working with these creatures three decades ago, I instantly experienced a sense of presence, a sense of familiarity, a pattern that connects. People describe swimming with dolphins as moving and even spiritual. What they are feeling, I think, is a reconnection with that primal pull, biophilia, in a powerful way. My three decades of rigorous study now places what I knew in my gut to be true into the realm of demonstrated science. The evidence of cognitive continuity is now, I believe, compelling, and because of that I feel compelled to act as a scientist on behalf of dolphins, to be their voice where they have none, to be their advocate.

I first stepped into a scientist-advocate role in the late 1980s, when the annual slaughter of more than 130,000 dolphins by tuna fishermen off the coast of San Diego came into public consciousness. Yellowfin tuna and dolphins have a special biological relationship that the fishermen exploited. Schools of tuna and several species of dolphin often swim together, for reasons that remain unknown. The fishermen used the presence of the dolphins at the water's surface as a signal that there might be tuna beneath. They then circled the dolphin with purse seine nets, hauled them up, and caught the tuna, and at the same time they caught hundreds of dolphins in which they had no interest. Many of the dolphins drowned while entangled in the nets; others were crushed in the heavy winch equipment used to haul the nets aboard the ship.

The California-based Earth Island Institute surreptitiously shot footage of the catches, and I saw it on television. It was gruesome. I spoke out on radio and television to support the efforts of the institute. Ultimately, a tremendous outcry from consumers along with crusading efforts by schoolchildren led to the 1990 decision of three major U.S. canneries to no longer buy tuna caught by encircling dolphins. The U.S. Congress passed the Dolphin Protection Consumer Information Act the same year, and canneries started to put DOLPHIN-SAFE TUNA labels on their products.

Today, Taiji is not the only place in the world that dolphins and whales face the darker side of *Homo sapiens*. Thanks to the efforts of the International Whaling Commission, whales are slaughtered far less now than they once were. Most of the eighty-eight global members of the IWC want a full moratorium on whaling. Today, dolphins and small whales continue to be killed in drive hunts in Japan, the Faroe Islands, and the Solomon Islands. Modern whaling occurs in Japan, Norway, and Iceland. A smaller number of whales are taken in subsistence hunting in Canada, Greenland, St. Vincent and the Grenadines, and Russia (Chukotka in Siberia).

Whales continue to be killed each year for commercial gain, sometimes under Japan's disingenuous claim of "scientific whaling." This term, coined in the late 1980s, is a loophole that allows the Japanese to kill whales in the name of science if they take their physical measurements, obtain genetic samples, and analyze the contents of their digestive tracts. It has been argued that the whales eat too much commercially important fish, and thus these competitors need culling — a specious argument based on selectively released data about the stomach contents found in some species.

The IWC offers no protection at all for small cetaceans — dolphins, porpoises, and small whales. Focusing on Taiji is not a numbers game for me. Bottlenose dolphins are not endangered as a species around the world. It is a question of man's inhumanity to a sentient species. It is the duty and obligation of scientists to speak out against it. If we can't stop what has been happening in Taiji, what hope do we have to apply our science to fixing anything?

I am looking for nothing less than a change in consciousness about ourselves and other animals, not in a fuzzy New Age way, but in a way based in science. I do not want a future generation to look back and say, "They knew what was happening. Why didn't they stop it?" I don't want to be part of a generation that, knowing what we do about these animals, allowed the drive hunts to continue.

Steven Wise, a legal scholar who teaches animal rights law at Harvard Law School and the Tufts University School of Veterinary Medicine, argues in his *Drawing the Line* that dolphins and the great apes should have protection under the law due to the fact that they demonstrate self-awareness and other high levels of intelligence. Thomas White, the Conrad N. Hilton professor and director of the Center for Ethics at Loyola Marymount University, in a book entitled *In Defense of Dolphins* goes even further. Since dolphins are, like humans, intelligent, self-aware beings with personalities, emotions, and the capability to govern their own behavior, he proposed they be viewed as "nonhuman persons" and valued and protected under the law as such. I worry about this argument, however — does it mean that other species may be mistreated? I take an animal *welfare* rather than an animal rights view — I would much prefer that all species are treated humanely and remain free from undue pain and suffering. In any event, it doesn't take a lot of complicated argument to make the case against the drive hunts. They simply must be stopped.

⌘

A question that I am often asked is whether dolphins should be kept in aquariums, or, to put it more graphically, kept in captivity. This is a very important issue, one that many people and animal rights organizations feel strongly about. The coalition against the drive hunts includes both colleagues from the zoo and aquarium community alongside animal welfare and animal rights activists. Yet the latter often oppose the former, arguing (as does *The Cove*'s central subject Ric O'Barry) that dolphins should under no circumstances ever be kept in captivity. The question is an important issue. But it is also not a simple, black-and-white issue. I wish it were.

What is black-and-white is the fact that dolphins have large and complex brains and use them in complex ways.

If we lived in a world in which there wasn't a single dolphin in an aquarium and we were deciding if we should have dolphins in aquariums, I would be among those loudly saying no. But we don't live in that world. We live in a world where there are dolphins in aquariums. So what, then, is the right thing to do?

I am absolutely opposed to dolphins being captured from the wild. Not a single dolphin should be taken from the wild and put into an aquarium. Aquariums in the United States, Europe, and Australia comply with this edict.

The second issue is if aquariums should continue to breed dolphins and keep those that are already there. Some people argue that captive dolphins should be released into the wild, which, while well-meaning, is naive and inhumane. Most of these animals have been born in aquariums; they've learned nothing about how to forage in natural environments, let alone how to defend against predators. Integrating into established social networks of wild populations could also be problematic. Releasing these animals into the wild would most likely present unenviable ends: starvation or shark bait. Those few remaining dolphins in aquariums that were taken from the wild (over twenty years ago) might not fare any better. Releasing these animals into the wild sets animal rights before animal welfare, to the detriment of the animals in question.

So what are the alternatives? Captive breeding programs could be discontinued and dolphins phased out by natural attrition. I've suggested that social groups of captive dolphins could be transfered to protected marine sanctuaries where they would be fed and cared for by trained staff.

In my mind, the only justification for aquaruims to mantain dolphins is if they become strong advocates for dolphins in political arenas. Given the fact that millions of people — families, students, educators — walk through the doors of aquariums each year, aquariums are uniquely positioned to provide transforming experiences about dolphins and to expand the number of advocates for the protection of dolphins and their ocean habitats. Aquariums have the power to educate children and adults about their responsibility to be stewards of the planet through active advocacy. It often takes a face-to-face experience with a dolphin in an aquarium to make that connection, to propel someone to take action. But sadly, that is often not enough. If a connection is made, aquariums need to further educate and guide their visitors into real advocacy.

For years, working at the Wildlife Conservation Society and being a member of the Animal Welfare Committee of the Association of Zoos and Aquariums, I have worked with many extraordinary scientists, wildlife veterinarians, conservationists, animal welfare advocates, and animal caregivers. Many have spent their lives trying to enrich the lives of the animals in their care, animals that effectively act as ambassadors for their wild counterparts so that they and their natural habitats are protected. Those captive ambassadors are real individuals with real needs. Is it not then an obligation for aquariums (and zoos) to be their advocates?

Aquariums, if they are to be viable in the future, must develop educational venues that allow people to experience for themselves the remarkable cognitive abilities and social prowess of dolphins. Rather than putting on 1950s-style dolphin shows, educational and entertaining films could be shown to teach visitors about dolphins, including the traits that have made them so interesting to the public from ancient times to the present. Aquariums need to provide experiences and tell stories that create the pattern that connects us to dolphins in a visceral way. And it is also the responsibility of aquariums to educate people about the plight of dolphins in the wild and engage their participation in alleviating the problems. In other words, I think that any aquarium that maintains social groups of dolphins must commit not only to attending to the welfare of the individuals in its care but also to fighting for the welfare of dolphins in the wild, including the conservation and protection of wild populations. If these two conditions are met, I support this second, more complex proposal, that aquariums maintain the current population of captive dolphins. I support it now in the world we live in, because at this point dolphins and whales need to be in the public eye and heart. We need to fight for their protection.

The past half a century has seen an odd contradiction in the evolution of zoos and aquariums, at least as far as dolphins are concerned. The first zoos were little more than menageries, often established by noblemen as curiosities, displaying exotic animals captured from the Dark Continent. The animals were viewed as spectacles or even freaks, housed in small cages with prisonlike bars. Beginning in the twentieth century, zoos gradually became wildlife parks, providing animals with more space and richer environments that approximated their natural habitats in some measure. No one is fooled into thinking that he is in a truly natural environment when he goes to a wildlife park, but from an animal welfare point of view, these establishments are a terrific improvement over menageries.

For dolphins in aquariums, the opposite has occurred. Where once they swam in the company of fish, turtles, seals, and other sea life, they are now too often housed in sterile tanks, in the interests of sanitation. Not all aquariums do this, of course, but most do. And most aquariums continue to see dolphin shows as not only appropriate but as centerpieces of their aquariums. Many managers of aquariums maintain that dolphin shows are what the public wants. They want to see dolphins do higher and higher jumps to reach the omnipresent ball hanging from above. Frankly, I find dolphin shows to be old style rather than forward looking and transformational. They take us back to the mentality of menageries, with animals being held as spectacles to be ogled. Having dolphins jump higher and higher and do ever more clever tricks demeans them as objects; it does not respect them for the kind of animals they really are. I view aquariums that indulge in these archaic displays as letting down both people and dolphins.

There once was a time when chimpanzees in zoos were dressed in human clothes and made to act out tea parties, ride bikes, smoke pipes, all kinds of activities that had nothing to do with the kind of creatures that chimpanzees are. One has to be getting on in years to remember these spectacles, because many years ago they were recognized as demeaning and stopped. When will making dolphins perform tricks that have nothing to do with their lives be similarly recognized as demeaning, and similarly stopped?

Millions of people go to aquariums each year because they love the animals they see. They also learn about the animals' minds, not just how big they grow, where they live, and what they eat. Given the chance, people are eager to learn more.

There's an exhibit at the Bronx Zoo that is a model for aquariums with dolphins. It's called the Congo gorilla forest, and it centers around a family group of gorillas that live in about as natural a setting as can be hoped for outside the mountains of Rwanda. Before visitors get to see the gorillas they are treated to an enthralling film in which a wildlife biologist talks about gorilla family life, social interactions, and foraging, and conjures up what it is like to be in the presence of gorillas. She also talks about the ugly side of gorilla life and the real threats they face, including poaching, which involves the beheading of gorillas and the removal of their hands and feet, prized trophies for misguided people somewhere in the world. The film ends with a young gorilla peering out at you through bushes as music rises in a crescendo — at this point you feel like you would do anything to protect these animals. It's theater, it's drama, it's reality, and viewers understand that they can do something that will make a difference. The screen rises and the curtains open, revealing an expansive green hillside with live gorillas. The visitor might be standing just a few feet away, though safely behind glass, from a mother nursing her young, or adolescents romping with one another on top of a patient silverback male. At the end of the exhibit, the visitor is allowed to choose which Congolese wildlife species he wants his entrance fee to help protect.

It is not a stretch to imagine an exhibit with dolphins that honors the same principle, that invites people into the lives of these animals, teaches them about the dolphins' plight in the world, and then gives them an opportunity to act. The National Aquarium, in Baltimore, where I currently do my research, is working toward a similar strategy, which is evident in its vision statement. It says that its goal is to "inspire our visitors and partners to celebrate and nurture the world's aquatic habitats from tropical rain forests to coral reefs; from our Chesapeake Bay to the world's oceans." It goes on to say that "through pioneering science, conservation, and educational programming, we will confront the pressing issues facing global aquatic habitats. Our legacy to future generations is a world in which aquatic habitats are preserved and restored." People can watch videos of dolphins interacting with the mirror, which is a very powerful way for them to understand that there is more to being a dolphin than doing physical tricks with a ball.

It isn't hard to imagine educational exhibits with dolphins that include active sessions with underwater keyboards. Such an interactive system would not only offer the dolphins more control in obtaining toys and activities but also present them in the context of animals that are "minded" individuals. It would serve as an interface between us and the dolphins so they could make observable and clear requests and we could comply. As I write these words, my colleagues and I are in the planning phase of an underwater touchscreen for dolphins. (The ones I used with Delphi and Pan are positively archaic by comparison.) And members of the public could be directly involved. Remember that Lou Herman taught his dolphins to understand and respond to short sentences of hand gestures. Visitors could watch as a dolphin is given sentences, even novel sentences that visitors create for the dolphins based on a set of known word-gestures. I am certain the audience would be amazed by their cognitive feats. Dolphins could demonstrate their natural sensory abilities too, such as their use of echolocation and its relationship to how they acoustically see the world. All it takes is some imagination and a leap beyond the idea that dolphins are little more than accomplished aquatic gymnasts with nothing going on in their minds.

❧

I have viewed my work with dolphins through two different lenses, distinct from each other but complementary. There is the lens of scientific rigor that I use to find that extra swath of evidence, that extra data point that will help prove a thesis. And there's the more lyrical lens, a lens that is sensitive to the patterns that connect all of us to that ineffable essence of nature. It is the part that responds emotionally to Eiseley's poem "Magic" when I read it to my students. It is the part that understands viscerally the presence of dolphins.

During the days when I was at Marine World, I was also on the faculty of San Francisco State University. One day I was walking across the campus escorted by Orson, one in a sequential line of three incredibly smart (and humongous) Newfoundland dogs that I have rescued and adopted over the years. A stranger walked up to us, looked directly into my dog's eyes, and said, "Wow, it really seems like there's someone in there." I share that very feeling when I am in the presence of dolphins. There is someone "in there" looking back at me. Not a person, but someone. An individual with a mind, not so unlike mine in many ways. There is also a sense of awe at their physicality — they are big, muscular, and supremely adapted to their environment. But there is more than that. There's a quality of intelligence. Despite profound evolutionary distances, despite profound differences in physical adaptations between us, one for terrestrial life, the other for life in the seas, we are profoundly similar in many cognitive domains. Yet at the same time, the dolphin is an alien intelligence and will perhaps never be seen clearly through our human lens.

The observation area for mirror work at the National Aquarium is a darkened shaft in the midst of a complex of three pools; there's a window into each pool onto which I can affix a two-way mirror so I can watch the dolphins' constantly inquisitive antics. It has to be completely dark inside this area we call "the pit" to transform a piece of acrylic into a mirror, so I cover the other windows with black velvet drapes. Despite the cramped quarters, I am usually joined by my collaborators Sue Hunter, the head of the animal programs at the aquarium, and one of my doctoral students, Rachel Morrison. We remain silent, not wanting the dolphins to be aware of us.

I watch little Foster cavorting playfully, opening his mouth and carefully inspecting inside or tilting his head and blowing a bubble, watching as it rises, and then doing it all over again. As I see these things I am aware of a mind at work that is far beyond that of my Newfoundlands, much as I loved them.

Humans and dolphins may be among the few species that reciprocate a very special emotion, empathy. I have been involved in many rescues of dolphins, some of which have been stranded on beaches. And I am always struck by the way that people rush to help. The impulse to do so is almost primal. There just is something about the presence of dolphins that touches us. At some point during the rescues I usually find myself standing in waist-deep water with at least one other person supporting the distressed animal so that it won't sink and drown, just as I stood with some of my students supporting Gordo in the hours before he died.

Not so long ago it struck me that what my colleagues and I do in these situations is mirrored by what dolphins sometimes do when they rescue a person in trouble at sea. They stop the person from sinking and drowning by positioning themselves alongside him, one on either side. What a beautiful symmetry of actions. Is there a beautiful symmetry of emotions here too? I do not know. But in the language of science, it seems to me more parsimonious to imagine that, to some degree at least, there is, rather than deny the possibility simply because I don't know.

The long loneliness that Eiseley spoke about was the idea that humans were special and apart from the rest of the living world, thanks to our sense of self, our capability of conscious altruistic behaviors born from empathy, our reflection and intention. And indeed, if we alone possessed these qualities, it would be a lonely planet. Yet each of these traits is shared with dolphins and some other species as well. We share more than we ever could have imagined with these other minds.

Eiseley wondered what intelligence would look like if we exchanged our hands for fins and our ungainly terrestrial forms for that of dolphins, these minds in the water. We would leave behind the trappings of civilization, where intelligence is reflected in what we do in the world, for good and for ill. Eiseley finished his essay reflecting on that hypothetical transformation, three sentences that I read as the final part of the final class of my animal-communication course. And I will end this book in the same way:

"Perhaps such a transformation would bring him once more

into that mood of childhood innocence in which he talked success-fully to all things living but had no power and no urge to harm. It is worth at least a wistful thought that someday the [dolphin] may talk to us and we to him. It would break, perhaps, the long loneli-ness that has made man a frequent terror and abomination even to himself."

Consortium of Marine Scientists and Zoo and Aquarium Professionals Call for an End to the Inhumane Dolphin Drives in Japan

(EXCERPTED FROM PRESS RELEASE ISSUED IN 2006)

An international consortium of scientists and leaders from the zoo and aquarium community are calling upon the Japanese government to end the infamous dolphin drive hunts in Japan. From September to April each year dolphin drive hunts are conducted by small groups of Japanese fishermen whose activities are regulated by the Japanese government. During the hunts, fishermen herd hundreds, sometimes thousands, of dolphins and other small cetaceans into shallow bays by banging on partially submerged rods that create a sonic barrier. Once there, the dolphins are corralled into nets and then speared, hooked, hoisted into the air by their tails by cranes in a manner that is inhumane by any standard. The animals are then slaughtered in an unjustifiably brutal manner, eviscerated alive, and many are left to die a long and painful death. The group contends that the hunts are a blatant violation of any reasonable animal welfare standards and are indefensible given the growing body of scientific research on dolphin cognitive abilities, cultural richness, and their capacity to experience pain and suffering.

Two years ago AZA and WAZA adopted advocacy statements calling for an end to the dolphin drives in Japan. A small delegation of zoo and aquarium scientists and professionals including Dr. Diana Reiss, Senior Research Scientist at the New York Aquarium of the Wildlife Conservation Society, Dr. Paul Boyle, Director of the New York Aquarium of the Wildlife Conservation Society, and Steve Olson, Director of Government Affairs of the AZA started a dialog and meetings with the Japanese Embassy to discuss the need to end dolphin drives. Despite compelling scientific evidence, these discussions have been unproductive.

This year, the delegation grew into a larger consortium of internationally renowned marine scientists and zoo and aquarium professionals comprised of Karen Sausman, President of the World Association of Zoos and Aquariums, Dr. Diana Reiss, Senior Research Scientist of the New York Aquarium of the Wildlife Conservation Society, Dr. Paul Boyle, Director of the New York Aquarium of the Wildlife Conservation Society, Dr. Lori Marino, Emory University, Dr. Sam H. Ridgway, University of California, San Diego, Dr. Louis M. Herman, University of Hawaii, Dr. Hal Whitehead, Dalhousie University, Dr. William E. Evans, University of Notre Dame, former chair of the U.S. Marine Mammal Commission, and Steve Olson, Director of Government Affairs for the AZA held a press conference on Wednesday, July 19, at National Press Club in Washington D.C. in which they reported scientific and ethical justification for ending dolphin drive hunts. Over 300 marine scientists have signed a statement saying that the hunts are an astonishingly cruel violation of any reasonable welfare standards and should end immediately. Numerous studies on the cognitive abilities, cultural richness, and above all, the capacity of dolphins to experience pain and suffering, mandate that the Japanese government should ban the hunts, which take place every year in the villages of Taiji and Futo. U.S. Senator Frank Lautenberg (D-NJ) also introduced a Congressional Resolution to the Senate last year condemning this practice.

The goal now is for a small delegation of marine scientists and zoo and aquarium professionals to meet with the Prime Minister of Japan and other government officials in Japan to provide them with the scientific evidence and ethical justification for ending the drives immediately. The consortium has started a global petition to gather one million signatures calling for the end to the dolphin drive hunts, which is listed on the Ocean Project Website at www .actfordolphins.org.

Notes

1. MINDS IN THE WATER

1 See Natalia Burns, ed., *Lore of the Dolphin* (Hillsboro, OR: Beyond Words Publishing, 2002), 7–10.
2 Ibid., 5–6.
3 See, for instance, Scott Taylor, *Souls in the Sea* (Berkeley, CA: Frog, Ltd., 2003).
4 See, for instance, Antony Alpers, *Dolphins: The Myth and the Mammal* (Boston: Houghton Mifflin, 1961), 6–9.
5 Ibid., 11–12.
6 Ibid., 13.
7 Ibid., 14.
8 Ibid., 15.
9 Ashley Montagu and John C. Lilly, *The Dolphin in History* (Los Angeles: University of California Press, 1963), 3–21.

2. FIRST INSIGHTS

1 John Lilly, *Lilly on Dolphins — Humans of the Sea* (New York: Anchor Press, 1975), vii.
2 Gregory Bateson, *Mind and Nature: A Necessary Unity* (New York: Dutton, 1979).
3 Diana Reiss, *Pragmatics of Human-Dolphin Communication* (PhD thesis, Temple University, 1983).

3. IN SEARCH OF THE DOLPHIN ROSETTA STONE

1 Marc Bekoff, *Minding Animals: Awareness, Emotions, and Heart* (New York: Oxford University Press, 2002), 47.
2 N. Wade, "Does Man Alone Have Language?" *Science* 208 (June 1980): 1349.
3 Diana Reiss, "The Dolphin: An Alien Intelligence," in *First Contact*, eds. Ben Bova and Byron Preiss (New York: NAL Books, 1990), 32–41.

4   Brenda McCowan and Diana Reiss, "Social Familiarity Influences Whistle Acoustic Structure in Adult Female Bottlenose Dolphins," *Aquatic Mammals* 24: 27–40.

5   Diana Reiss and Brenda McCowan, "Spontaneous Vocal Mimicry and Production by Bottlenose Dolphins (*Tursiops truncatus*): Evidence for Vocal Learning," *Journal of Comparative Psychology* 107 (1993): 301–12.

6   Ibid., 309.

4. NONTERRESTRIAL THINKERS

1   Reiss, "The Dolphin: An Alien Intelligence," 32–41.

2   Ken Marten et al., "Ring Bubbles of Dolphins," *Scientific American* (August 1996): 85.

3   Ibid., 86.

4   Ibid.

5   Brenda McCowan et al., "Bubble Ring Play of Bottlenose Dolphins (*Tursiops truncatus*): Implications for Cognition," *Journal of Comparative Psychology* 114 (2000): 98–106.

6   Rachel Smolker, *To Touch a Wild Dolphin* (New York: Anchor Books, 2002), 106.

7   Janet Mann et al., "Why Do Dolphins Carry Sponges?" *PLoS One* 3 (December 2008): 2.

8   Ibid., 3.

9   Michael Krützen et al., "Cultural Transmission of Tool Use in Bottlenose Dolphins," *PNAS* 102 (2005): 8939–43.

10   J. S. Lewis and W. W. Schroeder, "Mud Plume Feeding: A Unique Foraging Behavior of the Bottlenose Dolphin in the Florida Keys," *Gulf of Mexico Science* 1 (2003): 92–97.

11   R. G. Busnel, "Symbiotic Relationship between Man and Dolphins," *Transactions of the New York Academy of Sciences* 35 (1973): 112–31.

12   K. W. Pryor and J. Lindbergh, "A Dolphin-Human Fishing Cooperative in Brazil," *Marine Mammal Science* 6 (1990): 77–82.

13   C. K. Tayler and G. S. Saayman, "Imitative Behavior by Indian Ocean Bottlenose Dolphins in Captivity," *Behavior* 44 (1973): 286–98.

14   James Shreeve, "Machiavellian Monkeys," *Discover* (June 1991): 70; Stephen Jay Gould and Elizabeth S. Vrba, "Exaptation — A Missing Term in the Science of Form," *Paleobiology* 8 (1982): 4–15.

5. THE FACE IN THE MIRROR

1   Luke Rendell and Hal Whitehead, "Culture in Whales and Dolphins," *Behavioral and Brain Sciences* 24 (2001): 309–24.

2   Nathan W. Bailey and Marlene Zuk, "Same-Sex Sexual Behavior and Evolution," *Trends in Ecology and Evolution* 24 (2009): 439–46.

3   Gordon G. Gallup Jr., "Chimpanzees: Self-Recognition," *Science* 167 (1970): 86–87.

4   Charles Darwin, letter to S. E. Darwin, April 1, 1838.

5   Gallup, "Chimpanzees," 86.

6   Ibid., 87.

7   Gordon Gallup Jr., "Self-Awareness and the Emergence of Mind in Primates," *American Journal of Primatology* 2 (1982): 237–48.

8   Lori Marino, Diana Reiss, and Gordon Gallup Jr., "Mirror Self-Recognition in Bottlenose Dolphins: Implications for Comparative Investigations of Highly Dissimilar Species," in *Self-Awareness in Animals and Humans*, eds. S. T. Parker, R. Mitchell, and M. Boccia (Cambridge, UK: Cambridge University Press, 1994), 380–91.

9   Ken Marten and S. Psarakos, "Evidence of Self-Awareness in the Bottle-Nose Dolphin," in ibid., 361–79.

6. THROUGH THE LOOKING GLASS

1   Diana Reiss and Lori Marino, "Mirror Self-Recognition in the Bottlenose Dolphin: A Case of Cognitive Convergence," *PNAS* 98 (2001): 5942.

2   Ibid., 5937.

3   Philip Yam, "The Flipper Effect," *Scientific American* 285 (July 2001): 29.

7. COGNITIVE COUSINS

1   Tayler and Saayman, "Imitative Behavior by Indian Ocean Bottlenose Dolphins in Captivity," *Behavior* 44 (1973): 290.

2   Ibid., 291.

3   Andrew Whiten, "Imitation and Cultural Transmission in Apes and Cetaceans," *Behavioral and Brain Sciences* 24 (2001): 360.

4   W. R. Ashby, *An Introduction to Cybernetics* (London: Chapman and Hall, 1956).

5   Charles T. Snowdon, "Review of *Wild Minds: What Animals Really Think*, by Marc Hauser," *Natural History* (March 2000).

6   Douglas Adams, *The Hitchhiker's Guide to the Galaxy* (1979; repr., New York: Ballantine Books, 2009), 141.

7   H. Jerison, "Animal Intelligence as Encephalization," *Philosophical Transactions of the Royal Society of London, Biological Sciences* 308 (1985): 21–35.

8   Lori Marino, "A Comparison of Encephalization between Odontocete Cetaceans and Anthropoid Primates," *Brain, Behavior, and Evolution* 51 (1998): 230–38.

9   Suzana Herculano-Houzel, "The Human Brain in Numbers: A Linearly Scaled-Up Primate Brain," *Frontiers in Human Neuroscience* 3 (2009): 31, doi:10.3389/neuro.09.031.2009.

10   C. Butti et al., "Total Number and Volume of Von Economo Neurons in the Cerebral Cortex of Cetaceans," *Journal of Comparative Neurology* 515 (2009): 243–59.

11   Kenneth Oakley, *Man the Tool-Maker* (Chicago: University of Chicago Press, 1959).

12   Irven DeVore and Richard Lee, eds., *Man the Hunter* (Chicago: Aldine Publishing, 1969).

13   Nancy Tanner and Adrienne Zihlman, "Women in Evolution, Part I," *Signs: Journal of Women in Culture and Society* 1 (1976): 600.

14   Nicholas K. Humphrey, "The Social Function of Intellect," in *Growing Points in Ethology,* eds. P. P. G. Bateson and R. A. Hinde, (Cambridge, UK: Cambridge University Press, 1976), 303.

15   Ibid., 307.

16   Ibid.

17   Ibid., 309.

18   Ibid., 311.

19   Richard C. Connor, Jana J. Watson-Capps, William B. Sherwin, and Michael Krützen, "A New Level of Complexity in the Male Alliance Networks of Indian Ocean Bottlenose Dolphins (*Tursiops* sp.)," *Biology Letters* (November 3, 2010): doi:10.1098/rsbl.2010.0852.

20   Nicholas Humphrey, "Consciousness: A Just-So Story," *New Scientist* (August 1982): 475.

21   David Premack and Guy Woodruff, "Do Chimpanzees Have a Theory of Mind?" *Behavioral and Brain Sciences* 1 (1978): 515–26.

22   Louis Herman, "Exploring the Cognitive World of the Bottlenosed Dolphin," in *The Cognitive Animal,* eds. M. Bekoff et al., (Cambridge, MA: MIT Press, 2002), 275–83.

23   Joshua M. Plotnik et al., "Self-Recognition in an Asian Elephant," *PNAS* 103 (2006): 17053–57.

8. REFLECTIONS ON DOLPHIN MINDS

1   Virginia Morell, "Minds of Their Own," *National Geographic,* March 2008, p. 60.

2   M. C. Caldwell and D. K. Caldwell, "Epimeletic (Caregiving) Behavior in Cetacea," in *Whales, Dolphins, and Porpoises,* ed. K. S. Norris, (Berkeley, CA: University of California Press, 1966), 767.

3   Ibid.

4   Ibid., 773.

5   Ibid., 772.

6   "Dolphins Find Missing Sailor," *Cruising World,* March 1998, 10–11.

7   Maddalena Bearzi and Craig B. Stanford, *Beautiful Minds* (Cambridge, MA: Harvard University Press, 2008), 25.

8   Ibid., 26.

9   Gordon Gallup, "Can Animals Empathize? Yes," *Scientific American Presents* (Winter 1998): 68.

10 Frans de Waal, *The Age of Empathy* (New York: Three Rivers Press, 2009), 90.

11 Daniel Povinelli, "Can Animals Empathize? Maybe Not," *Scientific American Presents* (Winter 1998): 67.

12 Ibid., 75.

13 Peter Fimrite, "Daring Rescue of Whale off Farallones," *San Francisco Chronicle,* December 14, 2005.

9. INTO THE COVE

1   Boyd Harnell, "'Secret' Dolphin Slaughter Defies Protests," *Japan Times,* November 30, 2005.

2   Ibid.

3   Hiroko Tabuchi, "From Sea to Supermarket: Harrowing Look at Hunts," *New York Times,* October 23, 2009.

10. ENDING THE LONG LONELINESS

1   Loren Eiseley, "Magic," in *Notes of an Alchemist* (New York: Charles Scribner, 1972), 65–69.

2   E. O. Wilson, *Biophilia* (Cambridge, MA: Harvard University Press, 1984), 139.

3   Loren Eiseley, "The Long Loneliness," in *The Star Thrower* (New York: Harcourt, 1978), 37–44.

4   Ibid., 37.

5   Roger Lewin, "I Buzz Therefore I Think," *New Scientist* 1908 (January 15, 1994): 30.

6   Donald Griffin, "From Cognition to Consciousness," *Animal Cognition* 1 (1998): 5.

Acknowledgments

Writing this book has conjured up vivid memories of so many who have encouraged, helped, and supported me along the way. There are so many friends, colleagues and mentors to thank.

As a graduate student, sticking my foot into the waters for the first time, John Lilly and Betty Brothers gave me rare opportunities and I thank them for their generosity. My thanks to my graduate advisor Dennis Smith, whose conversations on communication theory and symbolism were so inspiring. I am deeply grateful to my mentors and dear friends Professors Rene-Guy and Marie-Claire Busnel for their endless encouragement and friendship and for teaching me about bioacoustics and French life.

Much of the work reported in this book would not have been possible without the amazing group of graduate, undergraduate students and volunteers who worked on Project Circe at Marine World. A special thanks to Brenda McCowan, Bill Baldwin, Laura Edenborough, Bruce Silverman, Denise Herzing, Cara Gubbins, Stacie Hooper, Spencer Lyn, and Jim Mullen and the marine mammal training staff. The lab would not have been possible without the support of Michael and Patty Demetrios, the Andersen Family Foundation, the Ampex Corporation, the Planetary Society, and Terry Kelly of the U S Geological Survey. I want to specifically acknowledge and thank the late Barney Oliver of the Hewlett Packard Corporation and the SETI program for his friendship and early support of Project Circe.

My lab at the Wildlife Conservation Society's New York Aquarium would not have been possible without the ongoing support and encouragement of many people. My sincerest thanks and admiration go to Paul Boyle (then director of the New York Aquarium) for his unwavering support of my research and his shared vision to save dolphins from the dolphin drives in Japan. My special thanks to Katie and Peter Dolan, Brian and Darlene Heidtke, and the City Council of New York, for their support of the research. I want to also thank Cynthia Reich, Richard Lattis and William Conway, for all of their help, support, and wise words over the years. A special thanks to Martha Hiatt and the rest of the marine mammal staff at the New York Aquarium for their assistance in the MSR study and who were part of the "larger team."

I also want to thank the management and staff of the National Aquarium in Baltimore especially Brent Whitaker, Deputy Executive Director for Biological Programs, and Sue Hunter, Director of Animal Programs.

I thank my many colleagues, collaborators, and graduate students over the years and today, who inspire me, challenge me and make doing research such an exciting adventure. In particular, I thank, Irene Pepperberg, Frans de Waal, Joshua Plotnik, Don Moore, Sue Hunter, Brent Whitaker, Hal Markowitz, Laurie Gage, and Peigin Barrett.

My deepest thanks to my extraordinary editor, Bruce Nichols, who has been an absolute joy to work with and from whom I have learned much about making a book. I also want to thank others on the Houghton Mifflin Harcourt team including Christina Morgan and Michelle Bonanno. I want to especially thank and acknowledge Roger Lewin for his material assistance with the draft manuscript. I am especially thankful to my agent, John Brockman, for his guidance, steering and support.

Above all I thank my inner circle: My parents Arthur and Jean, who told me that I could accomplish anything imaginable. My dolphin mentors, I thank each one of you for showing me so many very special things. My husband Stuart and my daughter Morgan, thank you for being there with me along the way.

Index

About the Author

Dr. Diana Reiss is a professor in the Department of Psychology at Hunter College and in the Biopsychology and Behavioral Neuroscience Program of the Graduate Center, City University of New York. She directs the Dolphin Research Program at the National Aquarium in Baltimore. She is also on the adjunct faculty of the Department of Ecology, Evolution, and Environmental Biology at Columbia University, and she served as a member of the Animal Welfare Committee of the Association of Zoos and Aquariums. Her research focuses on dolphin cognition and communication, comparative animal cognition, and the evolution of intelligence. She has authored papers published in numerous international scientific journals and book chapters, and her work has been featured in many television science programs.